SELECTED TOPICS ON OPTICAL AMPLIFIERS IN PRESENT SCENARIO

Edited by **Sisir Kumar Garai**

Selected Topics on Optical Amplifiers in Present Scenario
Edited by Sisir Kumar Garai

Published by InTech
Janeza Trdine 9, 51000 Rijeka, Croatia

Publishing Process Manager Maja Bozicevic

Technical Editor Teodora Smiljanic

Cover Designer InTech Design Team

Additional hard copies can be obtained from orders@intechopen.com

Selected Topics on Optical Amplifiers in Present Scenario, Edited by Sisir Kumar Garai
p. cm.
ISBN 978-953-51-0391-2

For EU product safety concerns:
IN TECH d.o.o., Prolaz Marije Krucifikse Kozulić 3, 51000 Rijeka, Croatia,
info@intechopen.com or visit our website at intechopen.com.

Meet the editor

Dr Sisir Kumar Garai was born in 1971 in West Bengal, India. He received the M. Sc. degree in Physics in 1994 and the M. Tech degree in Microwaves in 1998, and awarded the Ph. D degree in Science (Physics) on 6th February, 2012 from 'The University of Burdwan', West Bengal, India. His field of research interest includes Optoelectronics, Optical Computation and Communication, etc., and he has contributed more than twenty five papers in different International Journals in his area of research. He has received the best paper presentation award in '16th West Bengal State Science and Technology Congress-2009' and his name is also included in Marquis Who's Who-2010 & 2011, and in International Biography Centre (IBC)-Cambridge-2011. He has been an Assistant Professor and the -Head of the Department of Physics of M.U.C. Women's College, Burdwan, West Bengal, India, since 2002.

Contents

Preface

To fulfill the ever increasing demands of internet based communication network, the speed of computing as well as the speed of data processing should be high enough with the transmission medium of enormous potential bandwidth having the vast amount of information handling capabilities. On the other hand, conventional electronic technology has already reached its ultimate speed limit (40 Gb/s) through the limitations of miniaturization of chips and bandwidth limitation, and obviously it will create data traffic jam in future internet based networking services. Therefore, a drastic solution of the acute problem is needed and the scientists and technologists promote their thinking to a totally different track from the conventional electronic system so that the computer performance and data signal processing can be further improved with potential communicating medium to such an extent that we would be well prepared to accept the present and future challenges of data traffic.

Optical computing and optical signal processing are strongly believed to be the most feasible technology that can provide the way out of the extreme limitations imposed on the speed and the complexity of present days computation and communication by conventional electronics. Optics provides higher bandwidth than electronics, which enables more information to be carried simultaneously and data to be processed in parallel with impressive increase in speed by several orders of magnitude over that of the electronic signals. If the parallelism of optics be associated with fast switching speed of optical devices, it would result in the surprising computational speed and processing of data.

Considering the present scenario of speed and band width limitation of electronic computing, signal processing and future problem of data traffic, the scientists, technologists and researchers are working in the field of optical computing and optical signal processing in telecommunication network. The role of optical amplifier is indispensable in optical computation and optical data communication network. Over the past two decades, optical amplifiers such as Erbium doped fiber amplifier (EDFA), Raman Amplifier, Semiconductor Optical Amplifier (SOA) are the prime candidates as optical network functional components and have many functional applications such as wavelength conversion, regeneration, wavelength selection, booster, in-line amplification, in-node optical pre-amplification, and mid-span spectral inversion etc.

The selected topics in this book covers the roles of semiconductor optical amplifier (SOA) as the building blocks of the next generation of optical access network, high speed all-optical cascaded switches, frequency encoded all-optical logic processors; key element of optical packet switching architectures, multifunctional elements in microwave photonic system and use of it to generate tunable high power narrow-spectrum diode laser system for performing different advance functionalities in present scenario of optical communication network. Technology of upgrading the gain and noise figure of erbium doped fiber amplifier (EDFA) in shorter wavelength side and, the study of the variation of material gain of quantum dot (QD) structure over the long wavelength ranges are also included in this book

The book comprises eight chapters. The functionalities of SOA are spanned from chapter 1 to chapter 6. In chapter 1, the authors have presented the role of reflecting semiconductor optical amplifier in next generation of optical access network. Chapter 2 deals with a review work on successive development of SOA based optical switches regarding their speed limitation, signal to noise ratio and clearly mentioned the role of the turbo switch to overcome these limitation and finally illustrate the importance of cascaded- SOA based optical switches in optical signal processing. In chapter-3, the author has presented a method of developing all-optical frequency encoded logic processor exploiting the state of polarization rotation (SOP) character of the probe beam in nonlinear SOA. Here the author at first mentions the advantages of frequency encoded data over other conventional data encoding and then successively presents the method of generating frequency data, different logic gates and all-optical memory unit and finally mentions the way out of developing multivalued logic processors and application of the scheme in optical computing and WDM telecommunication network. The authors have presented the SOA based optical packet switching architectures in chapter-4. Here they have mentioned different switching paradigms and the superiority of SOA in optical packet switching and have established some optical packet architectures and illustrate their realization using SOA in elegant ways. The effectiveness of SOA in reducing the power consumption is also analyzed. The multi-functional capability of SOA in microwave photonic communication systems such as optical amplification with modulation, gating, photo-detection, dispersion compensation, linearization, etc. have been demonstrated in chapter-5. The chapter also describes the applications of SOA-modulator, SOA-detector and SOA-dispersion compensator in microwave photonic communication systems. Based on the performance of tapered semiconductor optical amplifier, the generation of three red tunable high-power narrow-spectrum diode laser systems is demonstrated in chapter-6 which has so many applications in optical communication network, like as pump source of different optical and optoelectronic devices. The chapter-7 covers the improvement of gain and noise figure of EDFA in shorter wavelength side using different macro-bending approaches and varying fibre parameters such as length, radius, etc. The knowledge of material gain of a medium is very important to design an optical amplifier. Therefore, study of the variation of material gain of quantum dot (QD) structure for p-type and n-type doping over the long wavelength (800-2300 nm) is included in chapter-8.

All the selected topics of this book are very interesting, well organized and the presentation is also very lucid. This book covers the emerging applications of optical amplifiers in present scenario and I believe that this book will be of great value not only to the researchers in the field of optical computing and data processing, optical telecommunications, but also to the component suppliers, postgraduate students, academics and anyone seeking to understand the trends of optical amplifiers in present scenario and the consequent changes in optical amplifier design and technology.

Without the unstinting support from so many persons, it would not have been possible for me to edit this book. Therefore, it is a great pleasure for me to take this opportunity to express my gratitude to all of them. First of all I would like to express my indebtedness to Aleksandar Lazinica, CEO of InTech Publisher for appointing me the Editor of this book. I am also grateful to all the writers for contributing their valuable research works in this book. Again I am indebted to Ms. Maja Bozicevic, Publishing Process Manager for her incessant help in numerous aspects to enable me to do the editorial work. I wish to convey my thanks to Technical Editorial staff and all other staff members of the InTech publisher. I am grateful to my respected teacher, Prof. Sourangshu Mukhopadhyay, University of Burdwan, India for his constant encouragement and valuable suggestions. Finally. I would like to extend my sincere thanks to all my colleagues for their incessant encouragement.

We shall deem our effort amply rewarded if the book wins the appreciation of the users.

Sisir Kumar Garai

Dr. Sisir Kumar Garai
Assistant Professor
M.U.C. Women's College, Burdwan
West Bengal,
India

1

Next Generation of Optical Access Network Based on Reflective-SOA

Guilhem de Valicourt
Alcatel-Lucent Bell Labs France, Route de Villejust, Nozay, France

1. Introduction

Communication networks have evolved in order to fulfil the growing demand of our bandwidth-hungry world. First, the coaxial cable has replaced the copper cable since 1950 for long- and medium-range communication networks. The Bit rate-distance product (BL) is commonly used as figure of merit for communication systems, where the B is the bit rate (bit/sec) and L is the repeater spacing (km). A suitable medium for transmission needed to be available and optical fibres were selected as the best option to guide the light (Kao & Hockham, 1966). A radical change occurred, the information was transmitted using pulses of light. Thus further increase in the BL product was possible using this new transmission medium because the physical mechanisms of the frequency-dependent losses are different for copper and optical fibres. The bit-rate was increased in the core network by the introduction of a new technique: Wavelength-Division Multiplexing (WDM). The use of WDM revolutionized the system capacity since 1992 and in 1996, they were used in the Atlantic and Pacific fibre optic cables (Otani et al., 1995).
While WDM techniques were mostly used in long-haul systems employing EDFA for online amplification, access networks were using more and more bandwidth. Access network includes the infrastructures used to connect the end users (Optical Network Unit - ONU) to one central office (CO). The CO is connected to the metropolitan or core network. The distance between the two network units is up to 20 km. The evolution of access network was very different from in the core network. High bit-rate transmissions are a recent need. At the beginning, it provided a maximum bandwidth of 3 kHz (digitised at 64 kbit/s) for voice transmission and was based on copper cable. Today, a wide range of services need to be carried by our access network and new technologies are introduced which allow flexible and high bandwidth connection. The access network evolution is obvious in Europe with the rapid growth of xDSL technologies (DSL: Digital Subscriber Loop). They enable a broadband connection over a copper cable and allow maintaining the telephone service for that user. In 2000, the maximum bit rate was around 512 kbit/s while today it is around 12 Mbit/s. However since 2005, new applications as video-on-demand need even more bandwidth and the xDSL technologies have reached their limits. The introduction of broadband access network based on FTTx (Fiber To The x) technology is necessary to answer to the recent explosive growth of the internet. Today, Internet service providers propose 100 Mbit/s using optical fibre. The experience from the core network evolution is a great benefit to access network. The use of WDM mature technology in access and

metropolitan network should offer more scalability and flexibility for the next generation of optical access network.
However the cost mainly drives the deployment of access network and remains the principal issue. Cost effective migration is needed and the cost capital expenditures (CAPEX) per customer has to be reduced. ONU directly impacts on the CAPEX. New optical devices are needed in order to obtain high performances and low cost ONU.
For uplink transmission systems using WDM, each ONU requires an optical source, such as a directly modulated laser (DML) (Lelarge et al., 2010). If wavelengths are to be dynamically allocated, one to each ONU, colourless devices are needed in order to minimize the deployment cost. Reflective Semiconductor Optical Amplifier (RSOA) devices can be used as a low-cost solution due to their wide optical bandwidth. The same type of RSOA can be used at different ONUs where they perform modulation and amplification functions. However, cost and compatibility with existing TDM-PONs is still an issue. As a consequence, hybrid (TDM+WDM) architecture is being investigated for next generation access network (An et al., 2004), as a transition from TDM to WDM PONs where some optical splitters could be re-used. Recently, the first commercial hybrid PON based on reflective semiconductor optical amplifiers (RSOA) has been announced (Lee H-H. et al., 2009). Such a network allows serving 1024 subscribers at 1.25 Gbit/s over 20 km.
In this chapter, the basic theory of SOA/RSOA is investigated. The different interactions of light and matter are described. Then, we focus on the device modelling. We develop a multi-section model in order to take in account the non-homogeneity of the carrier density. In this approach, we consider a forward and backward propagation as well as the amplified spontaneous emission (ASE) propagation. Longitudinal spatial hole burning (LSHB) strongly affects the average optical gain. An evaluation of the total gain in RSOA devices including the LSHB is proposed. The influence of the optical confinement and the length is described and leads to some design rules. Under the latter analysis, the performance of RSOA must be evaluated considering the trade-offs among the different parameters.
Dynamic analysis is then proposed in section 4. The RSOA modulation responses behave as a low-pass filter with a characteristic cut-off frequency. The carrier lifetime turns out to be a key parameter for high speed modulation and a decrease of its value appears to be required. The rates of recombination processes, such as stimulated, spontaneous and non-radiative recombination govern the carrier lifetime. They strongly depend on the position along the active zone and the operating conditions.
Furthermore, telecommunication networks based on RSOA are studied. We introduce the envisaged architectures of access networks based on RSOA. High gain RSOA is used as colourless transmitter and WDM operations are performed. Laser seeding configuration at 2.5 Gbit/s is realized and error free transmission is obtained for 36 dB of optical budget over 45 km of SMF. Low-chirp RSOAs enable a 100 km transmission at the same bit rate below the Forward Error Correction (FEC) limit. Direct 10 Gbit/s modulation is then realized using high speed RSOA.
Finally, we summarize the lessons learned in this chapter and conclude on RSOA devices as colourless optical transmitters.

2. Past history and basic concept of reflective SOA

In this section, the fundamental properties and the main concepts of SOA and RSOA are introduced in a simple way. We discuss the past history and the evolution of RSOA with the development of optical communication systems. Then, we detail the physics involved in SOA and RSOA. A multi-section approach is chosen to model the device behaviour under static conditions. The aim of this section is not to propose a complex model but to underline and understand the different mechanisms in RSOA devices.

2.1 RSOA evolution

The first idea of a Reflective SOA (RSOA) was proposed by Olsson in 1988 to reduce the polarization sensitivity via a double pass configuration using classic SOA.

The first integrated RSOA for optical communication appeared in 1996 where the device was employed for upstream signal modulation at a bit rate of 100 Mbit/s (Feuer et al., 1996). Then several experiments based on RSOAs for local access network were realized where the bit rate was increased to 155 Mbit/s (Buldawoo et al, 1997).

In this chapter, we mainly focus on the wavelength upgrade scenario for WDM-PON systems. While current optical access networks use one single upstream channel to transmit information, WDM systems use up to 32 channels increasing therefore the total transmitted information. RSOA is the perfect candidate because of its wide optical bandwidth, large gain and low cost, therefore fulfilling most of the requirements. For instance, its large optical bandwidth makes it a colourless cost-effective modulator for WDM-PON. The same type of device can be used in different ONUs, which reduces the network cost. Moreover, the large gain provided by an RSOA can compensate link losses without using an extra amplifier, which simplifies the overall solution.

Therefore since 2000, RSOA devices saw a fast growing interest in upstream channels transmission based on reflective ONU for WDM-PON applications. The first re-modulation scheme has been proposed in 2004 where the downlink signal was transmitted at 2.5 Gbit/s and uplink data stream was re-modulated on the same wavelength via the RSOA at 900 Mbit/s (Koponen et al, 2004). Then further investigations were realized and 1.25 Gbit/s re-modulation was demonstrated (Prat et al., 2005).

Today, several research groups work on these solutions such as Alcatel-Lucent Bell Labs, III-V Lab, Universitat Politècnica de Catalunya (UPC), Orange labs, ETRI, KAIST, IT and various optical devices manufacturers proposed commercial RSOA devices as CIP, MEL and Kamelian.

2.2 Fundamentals of SOA and RSOA

Gain in a semiconductor material results from current injection into the PIN structure. The relationship between the current I and the carrier density (n) is given by the rate-equation. The rate-equation should include the stimulated emission as well as the spontaneous and absorption rate. R(n) is the rate of carrier recombination including the spontaneous emission and excluding the stimulated emission. Electrons can recombine radiatively and non-radiatively therefore R(n) can be written as:

$$R(n) = R_{rad}(n) + R_{non-rad}(n) \tag{1}$$

When an electron from the Conduction Band (CB) recombines with a Valence Band (VB) hole and this process leads to the emission of a photon, it is called the radiative recombination. The rate of radiative recombination is:

$$\mathrm{R_{rad}(n) = B.\,n^2} \tag{2}$$

This term corresponds to the spontaneous emission recombination. Non-radiative processes deplete the carrier density population in the CB then fewer carriers remain available for the stimulated emission and the generated amount of light is limited. The three main non-radiative recombination mechanisms in semiconductor are:

- The linear recombination due to the transfer of the electron energy to the thermal energy (in the form of phonons). This mechanism is called the Shockley-Read-Hall (SHR) recombination. The rate of SRH recombination is:

$$R_{SRH} = A.n \tag{3}$$

- The Auger recombination due to the transfer of the energy from high-energy electron/hole to the low-energy electron/hole with subsequent energy transfer to the crystal lattice. The rate of the Auger transitions is:

$$R_{Aug} = C.n^3 \tag{4}$$

- Another non-radiative recombination process is the carrier leakage, where carriers leak across the SOA heterojunctions. The leakage rate depends on the drift or the diffusion of the carriers therefore is given by (Olshansky et al., 1984):

$$R_{leak} = D_{leak}.n^{3.5} \text{for diffusion and } R_{leak} = D_{leak}.n^{5.5} \text{ for drift} \tag{5}$$

The dominant leakage current is usually due to carrier drift. Therefore the total recombination rate is given by:

$$R(n) = R_{rad}(n) + R_{non-rad}(n) = A.n + B.n^2 + C.n^3 + D_{leak}.n^{5.5} \tag{6}$$

The carrier leakage is usually neglected. So the rate-equation states that the resulting change of the carrier density in the active zone is equal to the difference between the carrier supplied by electrical injection and the carrier's recombination. Amplification results from stimulated recombination of the electrons and holes due to the presence of photons. The interaction between photons and electrons inside the active region depends on the position and the time. Therefore the carrier density at z and t is governed by the final rate-equation. We neglect carrier diffusion in order to simplify the carrier density rate-equation. This assumption is valid as long as the amplifier length L is much longer than the diffusion length, which is typically on the order of microns. We also assume that the carrier density is independent of the lateral dimensions.

$$\frac{dn(z,t)}{dt} = \frac{I(t)}{e.V} - (A.n(z,t) + B.n(z,t)^2 + C.n(z,t)^3) - v_g g_{net}.S(z,t) \tag{7}$$

Where n(z,t) is the carrier density, I(t) is the applied bias current, S(z,t) is the photon density, g_{net} is the net gain and vg is the light velocity group.

A time domain model for reflective semiconductor optical amplifiers (RSOAs) was developed based on the carrier rate and wave propagation equations. The non linear gain saturation and the amplified spontaneous emission have been considered and implemented together in a current injected RSOA model (Liu et al., 2011). This approach follows the same analytical formalism as Connelly's static model (Connelly, 2007).

To make the model suitable for static analysis some assumptions have been made and simplifications have been introduced. Since, as a modulator, the RSOA is mainly illuminated by a CW optical source, the material gain is assumed to vary linearly with the carrier density but with no wavelength dependence. Amplified spontaneous emission power noise is assumed to be a white noise, with an equivalent optical noise bandwidth. When the current is modulated in a RSOA, the carrier density and the photon density are varying with time and position. This is caused by the optical wave propagation and the carrier/photon interaction. The carrier density variation is introduced in the model by dividing the total device length into smaller sections. For each section the carrier density is assumed to remain constant along the longitudinal direction. The equations are linking the driving current, the carrier density and the photon density. Figure 1 represents the model elementary section. It includes ports representing the input and output photon density (forward and backward), input and output amplified spontaneous emission (forward and backward), the input electrical current injection and carrier density. We do not consider the phase shift of the signal.

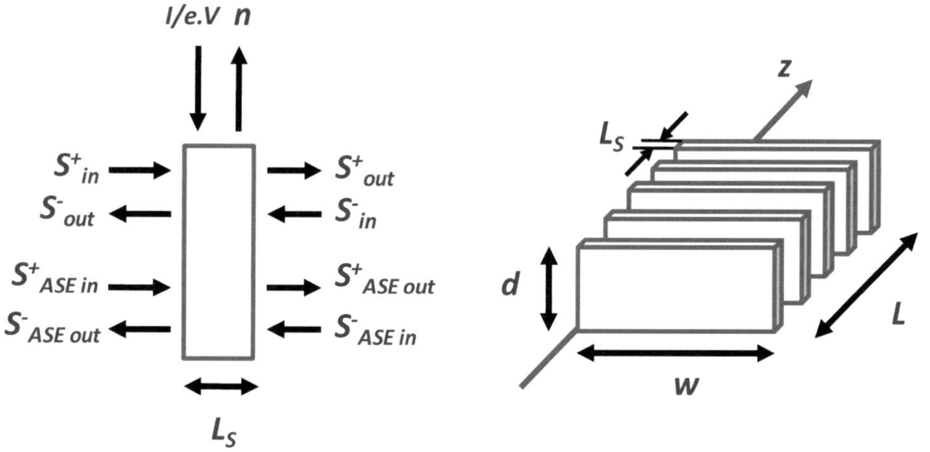

Fig. 1. RSOA elementary representation for numerical modelling

The forward and backward propagating optical fields (excluding spontaneous emission) are described by the relation between the input optical power and output optical power.

$$P^{\pm}(z \pm \Delta z) = P^{\pm}(z) e^{g_{net}\Delta z} \tag{8}$$

The material gain (g_m) is usually approximated by a linear function of the carrier density. In general the material gain also depends on the photon density S. For high photon density, the gain saturates and this phenomenon is described by the gain compression factor. Then, the material gain equation becomes:

$$g_S = \frac{g_m}{1+\varepsilon.S} \tag{9}$$

Where ε is the gain saturation parameter.
The boundary conditions for the device input and output facets, are given by:

$$\begin{aligned} P^+(0) &= (1 - R_1)P_{in} + R_1 P^-(0) \\ P^-(L) &= R_2 P^+(L) \end{aligned} \tag{10}$$

Where R_1 and R_2 are power reflection coefficients.

The amplified spontaneous emission is the main noise source in an RSOA and determines the RSOA static and dynamic performances under low input optical power. For high stimulated emission output power the spontaneous emission drops significantly and its impact on the device performances is less significant. For a section of length Δz the ASE power spectral density generated within that section is given by the following equation :

$$P_{ASE} = \eta_{Sp}(G_S(z) - 1)h\nu B_0 \tag{11}$$

Where G_s is the single pass gain of one section and η_{sp} is the spontaneous emission factor. The spontaneous emission factor can be approximated by (D'Alessandro et al., 2011):

$$\eta_{Sp} = \frac{n}{n - n_0} \tag{12}$$

In our model we have assumed a constant noise power spectral density over an optical bandwidth B_o. The bandwidth B_o is estimated at 5×10^{12} Hz. The implementation of the ASE noise travelling wave follows a procedure similar to the optical signal travelling wave. The spontaneous emission output power for the forward and backward noise signals has two contributions: the amplified input noise and the generated spontaneous emission component within the section. The gain variations with the bias current (Experimental and modelled) are compared in Figure 2.

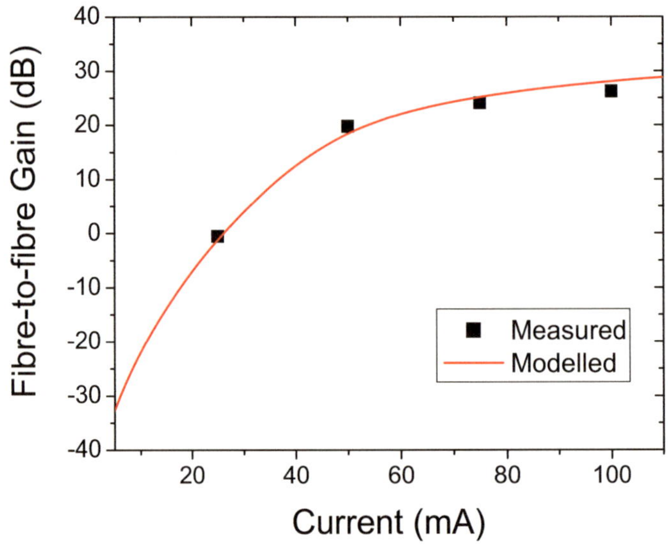

Fig. 2. Fibre-to-fibre gain for RSOA versus bias current

The carrier density profile is represented in figure 3 as well as the total photon density. At low input injection (P_{in} = -40 dBm), the carrier density profile is in this case not symmetrical due to the high reflection of the second facet. Strong depletion occurs from the ASE and the signal double propagation (reflective behaviour of the device). Also at P_{in} = -40 dBm, the ASE power dominates the signal power which explains that the RSOA device saturates more at high input electrical current.

At high input optical power, the carrier density in an RSOA is flattened due to the forward and backward propagations of the signal inside the device. The saturation effect occurs all along the RSOA. At the mirror and input facets, the signal photon density becomes larger with the injected current as it has been more amplified during the forward and backward propagations.

From this preliminary analysis, a general conclusion can be deduced. RSOAs should saturate faster than classic SOAs. The overall photon density inside a RSOA is larger than in a classic SOA, reducing the material gain available for signal amplification. However the forward and backward signal amplifications could compensate for this effect. Large photon density should also affect the E/O bandwidth and could be useful to obtain high speed devices. All these effects are stronger at high input electrical injection and high input optical power.

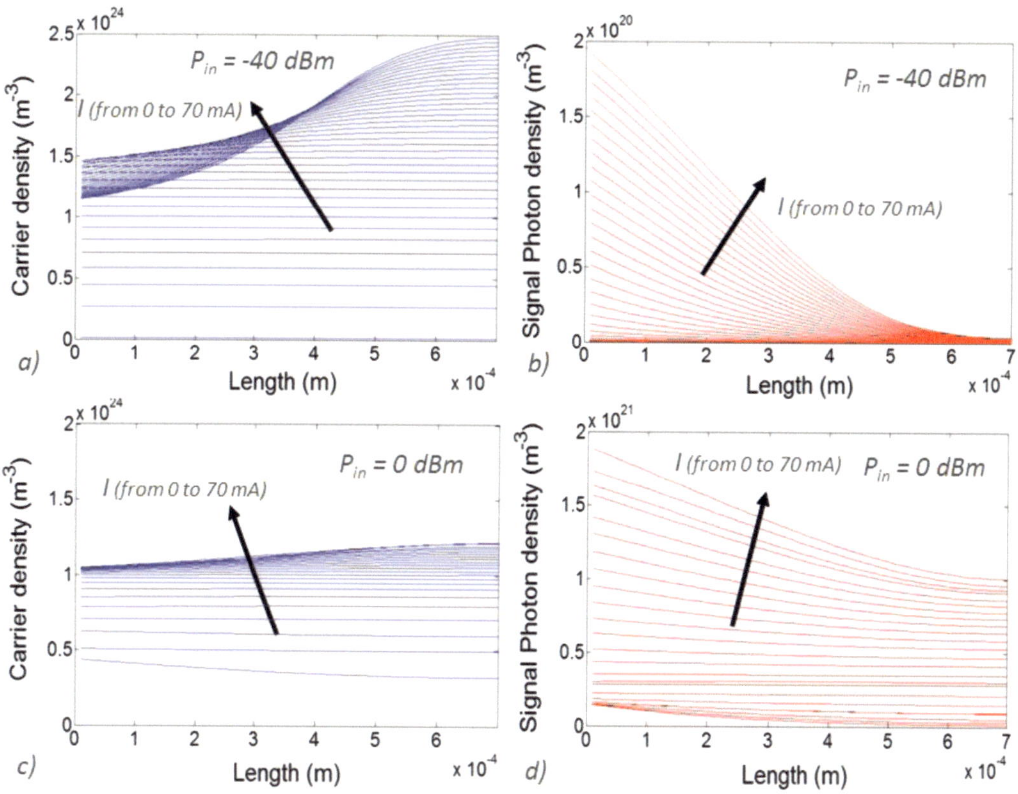

Fig. 3. Carrier and photon density spatial distribution in RSOA device. (a) and (b) represents the simulation from Pin = -40 dBm ; (c) and (d) for 0 dBm

3. RSOA devices static characteristics

Optical gain measurements depending on the input current and optical power were realized. Figure 4 shows the experimental setup which is used to perform static measurements. The required wavelength controlled by an external cavity laser is launched into the RSOA through an optical circulator (OC). A combined power meter and attenuator is used to control the input power to the RSOA. An optical spectrum analyser and a power meter are used in order to determine the static performances of the device, such as optical gain, gain peak, bandwidth and ripple, noise figure and output saturation power. The impacts of these several parameters (Γ and L) are experimentally studied in the next sub-sections.

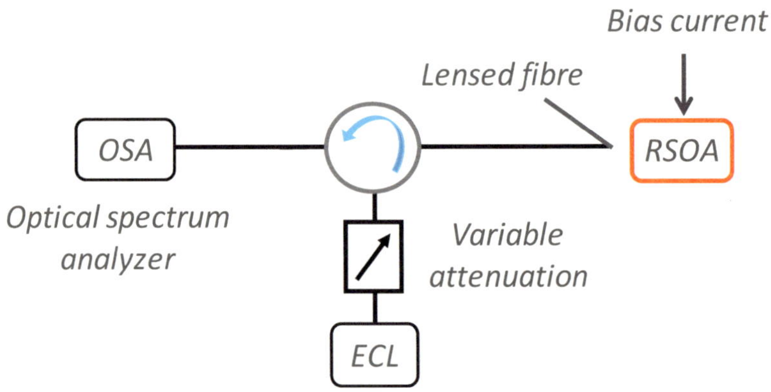

Fig. 4. Static experimental setup

3.1 Influence of the optical confinement

The optical gain for the optical confinements of Γ~20% and Γ~80% are compared depending on the input electrical current and optical power in Figure 5. The gain increases with the bias current as modelled in previous section and starts to saturate at high electrical injection. The low confinement factor (Γ~20%) devices show higher gain than the high confinement factor (Γ~80%) devices. This result is counter-intuitive as the net gain should increase with higher optical confinement and therefore the single pass gain. High Γ means more ASE and more saturation. Thus a low confinement factor induces lower spontaneous emission power by reducing the effect of the ASE inside the device (Brenot et al., 2005). As the RSOA is less saturated, the single pass gain is also increasing with the reduced confinement factor (because the LSHB is reduced).We demonstrated that RSOA devices have a non-uniform carrier density along the active zone. This interpretation can be confirmed by the simulations of the carrier density spatial distribution (section 2) and SE measurements (section 3.2).

Increasing the input power, the gain drops quickly due to the saturation effect. That is, the increase of optical input power at a constant current consumes many carriers for the stimulated emission therefore decreases the carrier density and increases the saturation effect. This transition corresponds to the frontier between the linear and the saturated regime. In this regime, the noise factor increases due to gain saturation. A common and useful figure of merit is the dependence of the optical gain on the output power. From this curve, we obtained the saturation power (P_{sat}) when the gain drops by 3 dB. Figure 5 (b) shows the optical gain versus the output power.

Most of SOA devices show saturation power around 10 dBm and optimized SOA can reach 20 dBm (Tanaka et al., 2006). However optimizing for maximum saturation power induces low gain (<15 dB) and large energy consumption (I > 500mA). In RSOA devices, high gain is obtained as well as reasonable saturation power.

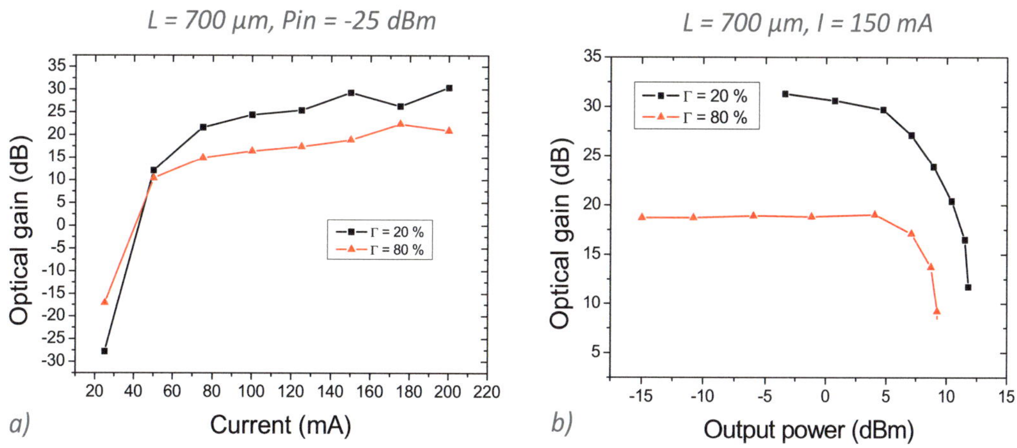

Fig. 5. Confinement effect on 700 µm long RSOA depending on the current (a) and the output power (b)

3.2 Saturation effect in long RSOA

Two optical confinement values have been studied and low optical confinement (Γ~20 %) enables the fabrication of high gain devices. It was the consequence of the LSHB reduction inside the active material which leads to an overall higher gain. However, the length (L) also affects the single pass gain (G_s). Again two effects are in competition inside the active zone: the exponential growth of G_s with the length and the non-homogenous carrier density distribution (which leads to strong saturation effect). Therefore a trade-off needs to be found in order to balance these two effects. By increasing the length, the forward and backward amplifications are also increased up to an optimum point. Devices that are too long induce high saturation and reduce the optical gain. Figure 6 (a) shows the optical gain versus the current density in different RSOA cavity lengths. The current density (J) is more relevant from a device point of view in order to compare similar operating conditions.

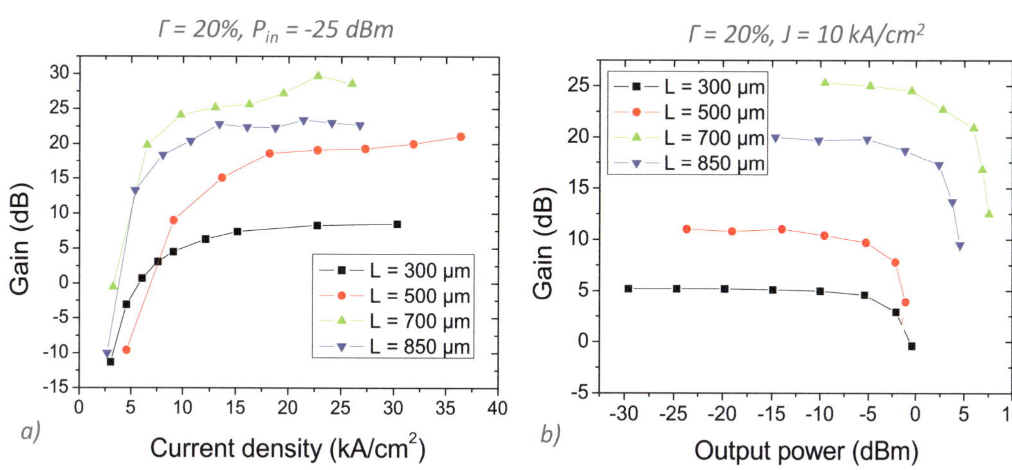

Fig. 6. Length effect on 20% optical confinement RSOA depending on the current density (a) and on the output power (b) for J = 10 kA/cm2

At first, the increase of the cavity length induces higher optical gain (from 300 μm to 700μm) however when it reaches 850 μm, the gain drops back. Therefore a maximum gain is obtained for 700 μm long devices. The optical gain versus the output power is presented in Fig. 6. (b) at the current density J = 10 kA/cm2. We can notice that increasing the gain leads to higher saturation power. It can be explained by the fact that we are at a constant current density therefore the electrical bias current increases with the length of the device leading to an improvement of the saturation power. For one specific optical confinement (Γ = 20%), an optimal length can be found in order to obtain the best static performances (high optical gain). At first, the optical gain increases linearly with the length. In fact, the forward and backward amplifications control the single pass gain. Figure 7. (a) represents the SE measurements where an optical fibre is placed along the active zone at the input/output, centre and mirror region. Then SE measurements as a function of the injected current are measured. SE measurements are performed in 700 μm long RSOA in order to confirm the presence of the saturation effect.

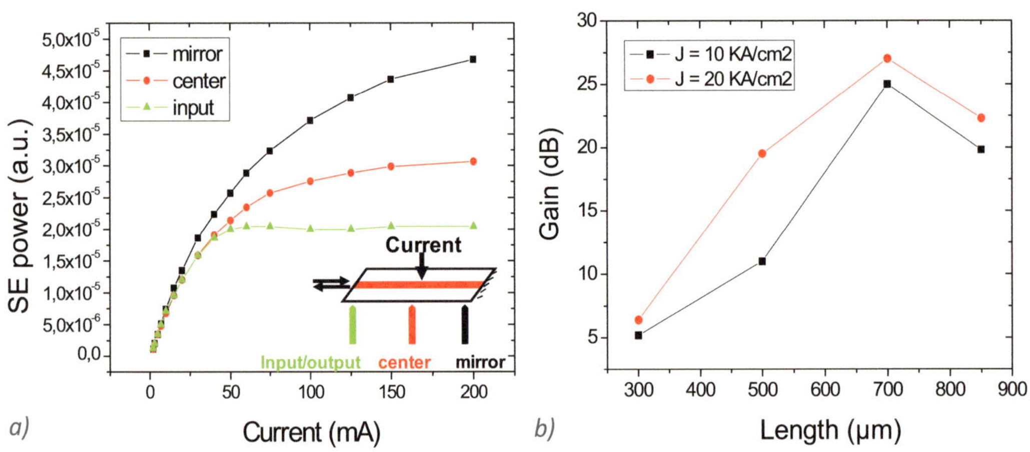

Fig. 7. (a) SE schematic and measurements; LSHB effect on (b) the optical gain in RSOA device

At low input bias current, no difference is observed due to the flat carrier density. The saturation effect starts to appear above 50 mA when the carrier density spatial distribution becomes non-homogeneous. Low SE power is collected at the input region due to the saturation effect which means low carrier density in the region. However the mirror region emits more SE power due to the high carrier density value. This demonstrates the presence of a strong saturation effect in the device.

In longer RSOAs, the depletion becomes stronger which induces a lower overall carrier density and a larger absolute difference in the carrier density between input and mirror facet. When varying the length of the RSOA, those several effects account for the existence of an optimum length where the optical gain is maximised. The optical gain versus the length of the device is plotted on Figure 7. (b) for two current densities.

4. Modulation characteristics and performances

RSOA devices have limited electro-optical (E/O) bandwidth between 1 to 2 GHz (Omella et al., 2008) compared to laser devices usually between 8 to 10 GHz. The difference can be explained by two effects that are not present in RSOA devices. The first effect is gain clamping. The carrier density stays low even at high electrical input current while the photon density is increasing. This produces a shorter carrier lifetime particularly advantageous for high speed modulation. The second effect is the electron to photon resonance due to the presence of a cavity. The resonance appears in the modulation response increasing the effective -3dB E/O bandwidth.

The absence of cavity in RSOAs limits the modulation speed of this device. The modulation response behaves as a low pass filter with a characteristic cut-off frequency (when the link gain drops by 3dB). One limitation is due to carrier density spatial distribution. High carrier density combined with low photon density induces long carrier lifetime. Furthermore the carrier and photon densities strongly depend on the position z along the device. Therefore a non-homogeneous carrier lifetime is obtained.

4.1 Carrier lifetime analysis

The objective is to obtain a first order approximation of the carrier lifetime for the steady state condition. We can demonstrate that the carrier lifetime can be approximated by:

$$\frac{1}{\tau_{eff}} = A + B.n + C.n^2 + \Gamma \times a \times S \times v_g \qquad (13)$$

Where the differential gain is defined by $a = \frac{\partial g}{\partial N}$, Γ is the optical confinement factor and S is the total photon density including the signal and the ASE.

The carrier lifetime is inversely proportional to the recombination rate. The recombination rate can be described using two different terms: one directly proportional to the spontaneous emission and non-radiative recombination (due to the defect or Auger process as described in section 2.2) and the second one depending on the stimulated recombination.

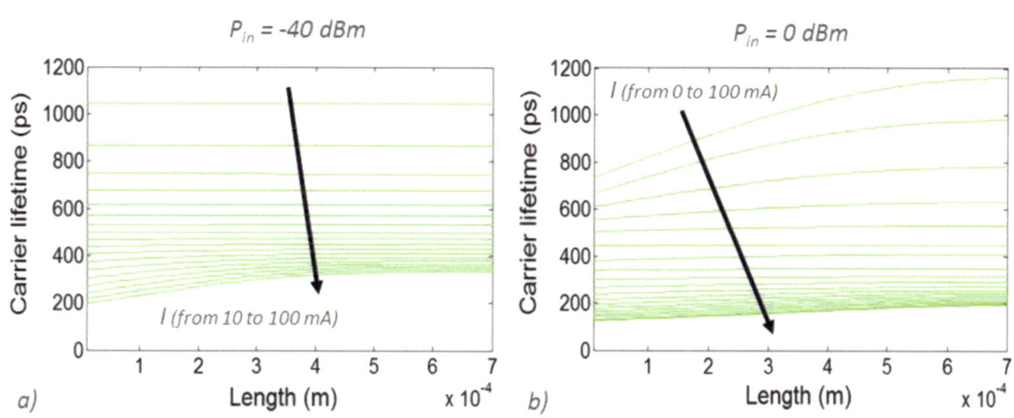

Fig. 8. Carrier lifetime simulation along 700 μm RSOA device at (a) low (Pin = -40 dBm) and (b) high (Pin = 0 dBm) optical injection

Simulations of the carrier lifetime have been carried out along the active region. Figure 8 represents the results with the bias current as parameter at P_{in} = -40 dBm (Figure 8 (a)) and Pin = 0 dBm (Figure 8 (b)). Obviously, in both cases, carrier lifetime decreases by increasing the input electrical current. It is mainly due to the increase in all recombination terms. The second important observation is the non-uniformity of the carrier lifetime along the device. At large optical input power (P_{in} = 0 dBm), the saturation effect described in section 3.2 is much stronger than with low input injection at low bias current. The average carrier lifetime is also smaller in this condition, due to a larger photon density. In order to understand the influence of the different recombination mechanisms on the carrier lifetime, it is important to follow the evolution of the different recombination terms depending on the bias current and the input optical power.

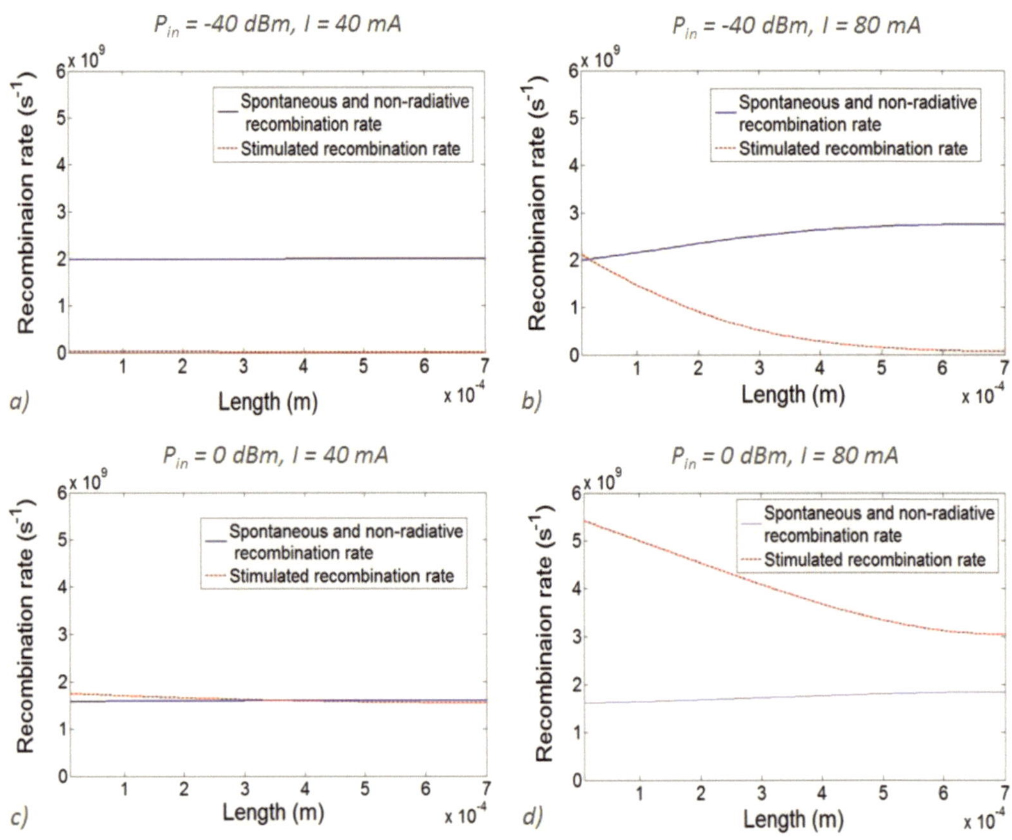

Fig. 9. Spatial distribution of spontaneous and non-radiative recombination rate compared to stimulated recombination rate in 700 μm long RSOA at different input conditions. (a) Pin = -40 dBm and I = 40 mA, (b) Pin = -40 dBm and I = 40 mA, (c) Pin = 0 dBm and I = 40 mA and (d) Pin = 0 dBm and I = 80 mA

Figure 9 represents the spatial distribution of the two terms at various operation conditions. At low input optical power ((a) and (b)), the spontaneous and non-radiative recombination rates are dominant even at high bias current. Therefore the carrier lifetime depends on this recombination term. At high input optical power ((c) and (d)), the photon density is much higher than in the previous situation, thus the stimulated recombination rate tends to overcome the spontaneous and non-radiative recombination terms. This is also confirmed at high input bias current (d) when the signal and ASE are strongly amplified along the RSOA.

However at low bias current (c), both phenomena balance each other and both are responsible for the carrier lifetime. They are more or less equal and do not vary that much over z. This analysis is crucial for digital modulation as the input conditions change over time, therefore the dynamic of the device will depend on which recombination rate is dominant at a precise time.

In order to validate our simulation, a comparison with experimental measurements should be done. High-frequency characterization is then needed. The experimental set-up and results are described in the next section.

4.2 High-frequency experimental set-up and characterization

We realize a RSOA-based microwave fibre-optic link as depicted in figure 10. All different devices of this experimental set up can be considered as two-port components and classified according to the type of signal present at the input and output ports. E/E, E/O, O/E or O/O are possible classifications where an electrical (E) signal or an optical (O) signal power are modulated at microwave frequencies (Iezekiel et al., 2000).The RSOA is considered as an E/O two-port device which is characterized by the electro-optic conversion process, i.e. the conversion of microwave current to modulated optical power.

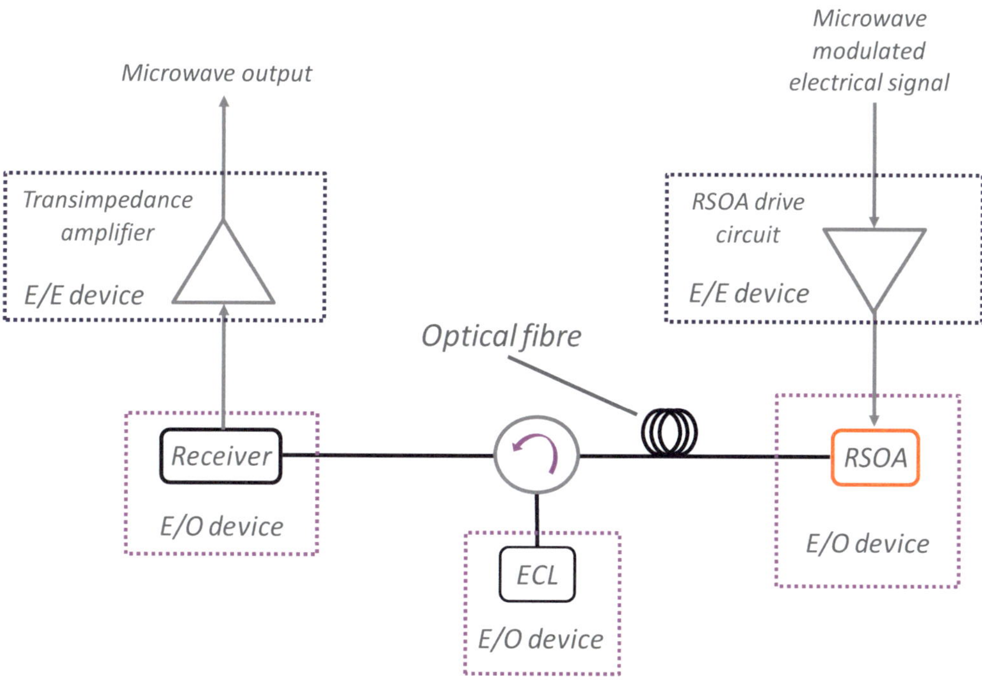

Fig. 10. High speed fibre-optic link

A full two-port optical characterisation of the complete set up is important to quantify the system performances. Dynamic characterization allows the measurement of the electrical response of the two-port network. A high-frequency signal is sent to the RSOA and the optical modulation is detected by a photodiode. The $|S_{21}|^2$ parameter (link gain) is measured over a range of frequency from 0 to 10 GHz. Figure 11 shows the electrical response of a typical RSOA device.

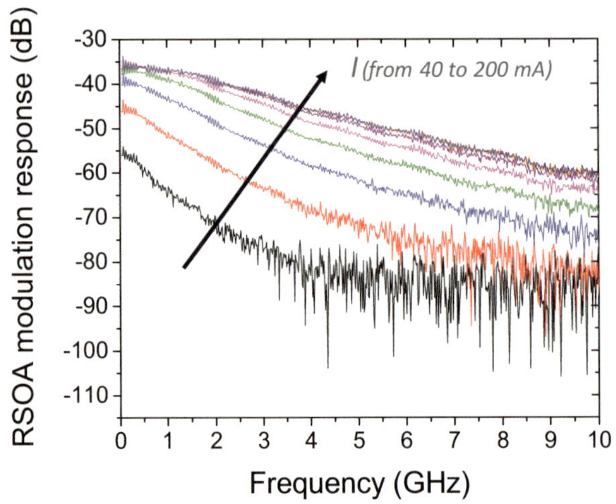

Fig. 11. Direct modulation measurements S21 in 700μm long RSOA device

We simulate the modulation bandwidth depending on the carrier lifetime based on the first order approximation. The carrier lifetime can be estimated along the RSOA but shows a non-homogenous spatial distribution. The first approach consists of considering an average carrier lifetime over the whole device. Simulation and experimental data are compared in Figure 12-(a) for a 700 μm long RSOA at 80 mA. The simulation results fit well with the measurements over a limited range (from 0 to 2GHz). The difference beyond can be explained by the addition of the buried ridge structure (BRS) limitation. In fact, the BRS equivalent electrical circuit exhibits a cut-off frequency around 3 GHz.

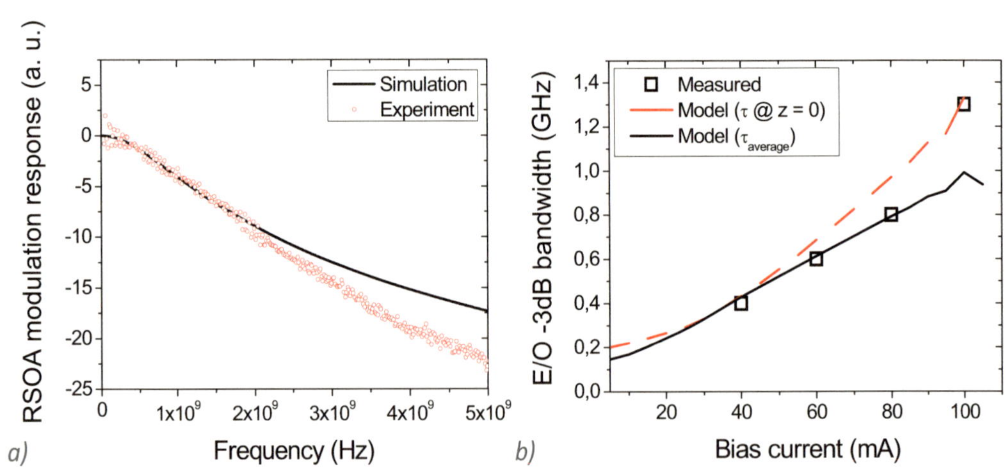

Fig. 12. RSOA (a) E/O modulation bandwidth versus frequency at I = 80 mA (b) -3 dB E/O modulation bandwidth versus bias current for 700μm of AZ

The -3 dB E/O bandwidth has been extracted from Figure 11 and plotted in Figure 12-(b). A second approach is proposed by simulating the modulation bandwidth based on the carrier lifetime at z = 0 where the saturation effect is stronger. At low bias current, the first approach fits better with the experimental values. However at high electrical current (I > 80mA), the second model is more adapted.

The simulations confirmed by the measurements describe why the modulation bandwidth is limited in RSOA devices. It is mainly due to a larger carrier lifetime than in laser which is caused by a smaller photon density. The effective carrier lifetime depends on several recombination rates and strongly on the operating conditions. The stimulated recombination rate can be increased at high input optical power and electrical current. These conditions induce high photon density inside the active zone reducing the carrier lifetime and increasing the -3 dB E/O bandwidth. However these conditions are not suitable for low power consumption networks. Therefore another solution for increasing the photon density seems to be a required condition to push back the RSOA frontiers. A 3 GHz modulation bandwidth can be obtained with 850 μm long RSOA, which has led us to the first eye-opening of a RSOA at 10 Gbit/s without electrical equalization or strong optical injection. More details are presented in section 6.2.

5. System performances

The role of a RSOA as an optical transmitter is to launch a modulated optical signal into an optical fiber communication network. Reflective semiconductor optical amplifier (RSOA) devices have been developed as remote modulators for optical access networks during the past few years and their large optical bandwidth (colorless operation) has placed them in a leading position for the next generation of transmitters in WDM systems. In RSOA devices, the wavelength is externally fixed. Various options have been studied such as using multi-wavelength sources (such as tuneable lasers, External cavity laser (ECL) , Photonic Integrated Circuits (PIC) or a set of Directly Modulated Laser (DML) at selected wavelengths), creating a cavity with the active medium of the RSOA, or using filtered white source. Therefore, RSOA devices as colourless transmitters can be used in different configurations:

- Laser seeding
- Spectrum-sliced EDFA seeding
- Wavelength re-use
- Self-seeding

In the laser seeding approach, the multi-wavelength external laser source can be located at the CO (Chanclou et al., 2007) or at the remote node (de Valicourt et al., 2009). From the CO, the optical budget is limited to 25 dB and strong RBS impairments appear. These limits are overcome by locating the laser at the remote node. One laser per remote node is needed, thus raising deployment cost, control management and power consumption issues.

Another possible architecture is using spectrum-sliced EDFA seeding. An erbium-doped fibre amplifier (EDFA) is used as a broadband source of un-polarised amplified spontaneous emission and this broad spectrum is then sliced by the Arrayed Waveguide Grating (AWG) for each ONU (Healey et al., 2001).

Wavelength re-use has been developed by Korean and Japanese companies (Lee W. R. et al., 2005). The downstream source from the CO is re-modulated as an upstream signal at the

ONU using RSOA. Simple efficient ONU is obtained as no additional optical source is needed.

The final approach is using a RSOA-based self-seeding architecture. This recent concept has been proposed by Wong in 2007. This novel scheme uses at the remote node (RN) a reflective path to send back the ASE (sliced by the AWG) into the active medium. The self-seeding of the RSOA creates a several km long cavity between ONU and RN. The wavelength is determined by the connection at the RN. This technique is attractive because a self-seeded source is functionally equivalent to a tuneable laser. Recent progresses show 2.5 Gbit/s operation based on stable self-seeding of RSOA (Marazzi et al, 2011). Another way to obtain self-seeding configuration is using an external cavity laser based on a RSOA and Fiber-Bragg Grating (FBG). BER measurements show that the device can be used for upstream bit rates of 1.25 Gbit/s and 2.5 Gbit/s (Trung Le et al., 2011).

In this chapter, we focus on the laser seeding approach. We present the scheme of a laser seeding architecture based on RSOA on Figure 13. Actually, Figure 13 shows the up-stream part of the link using an RSOA, i.e. the information sent from the subscriber to operator/network. At the central office, a transmitter is used to send light (containing no information) to the subscriber through an optical circulator. Light propagates through several kilometres of optical fibre. The signal is then amplified and modulated by the RSOA in order to transmit the subscriber data for uplink transmission.

In this section, we demonstrate an extended reach hybrid PON, based on a very high gain RSOA operating at 2.5 Gbit/s. To reduce RBS impairments, we locate the continuous wave (CW) feeding light source in the remote node, and the large gain of the RSOA allows using moderate CW powers. Alternative devices such as remotely pumped erbium doped fiber amplifier (EDFA) can be used in order to avoid the deployment of active devices in a remote node; this approach could also reduce the RBS level owing to a lower seed power and the management cost of the system (Oh et al., 2007).

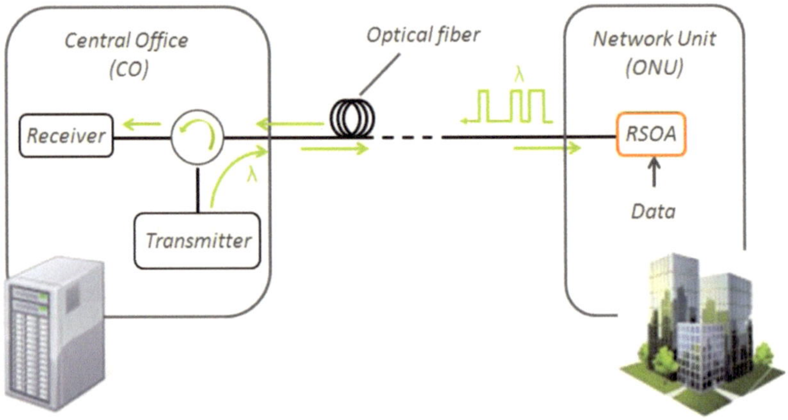

Fig. 13. Laser seeding Network architecture based on RSOA

Figure 14 shows the up-stream part of the proposed link. At the remote node, an external cavity laser (ECL) is used to launch an 8 dBm CW signal into the system through an optical circulator (OC). A wavelength demultiplexer is used to break a potential multi-wavelength signal back into individual signals. A given wavelength represents one of up to 8 sub-PON on a 100 GHz grid (from λ_1 = 1553.3 nm to λ_8 = 1558.9 nm). The output of the wavelength demultiplexer is coupled into a 20 km long Single Mode Fibre (SMF) followed by a 12 dB optical attenuator used to simulate a passive splitter for 16 subscribers. The CW signal is then modulated by the RSOA, generating the upstream signal. The RSOA is driven by a 2^{31}-1 pseudo-random bit sequence (PRBS) at 2.5 Gbit/s, with a DC bias of 90 mA. From the remote node, the upstream signal propagates on another 25 km long SMF which simulates the reach extension provided by the proposed network design. A variable optical attenuator is placed in front of the receiver in order to analyze the performance of the system as a function of the optical budget. This attenuator also accounts for the insertion-loss of the multiplexer at the CO (between 3 to 5 dB). Bit-error-rate (BER) measurements are done using an Avalanche Photo-Diode (APD) receiver and an error analyzer.

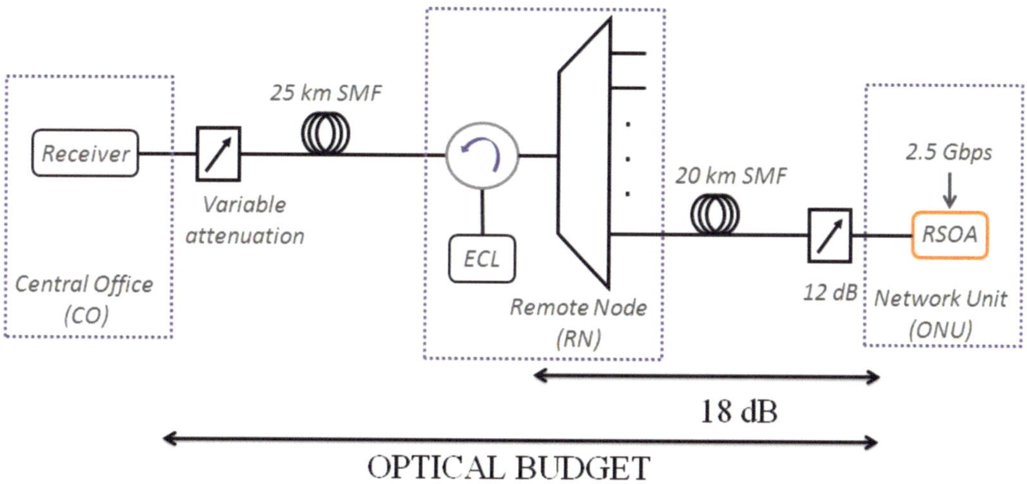

Fig. 14. Experimental setup of WDM/TDM architecture using RSOA (de Valicourt et al., 2010a)

At low bit rate, the best trade-off between gain, modulation bandwidth and saturation power is obtained for a 700 µm long cavity RSOA, therefore we chose this device in the experimental setup. The RSOA is driven at 90 mA with a -10 dBm input power. Figure 15 displays BER measurements performed at 1554.1 nm and 2.5 Gbit/s as a function of the optical budget between the CO and the extended optical network unit (ONU). The inset shows the open eye diagram measured at the output of the RSOA. Sensitivities at 10^{-9} in back-to-back (BtB) configuration and after transmission are -32 dBm and -27 dBm respectively. These performances are mainly due to the large output power of the RSOA, which allows for an increased optical budget compared to standard RSOAs: a BER of 10^{-9} is thus measured with an optical budget of more than 36 dB. Whatever the OB, the input power in the RSOA is -10dBm, which ensures that the device operates in the saturated regime, with a reflection gain of 20 dB. Gain saturation leads to a low sensitivity of the RSOA to back-reflections, since the output power only slightly depends on the input power. In Figure 15 (b), the BER of 8 WDM channels (100GHz spacing), is shown, for a 40 dB optical budget ; in this case, the BER is 10^{-7}, well below the forward error correction (FEC) limit. No penalty is observed due to the large bandwidth of the RSOA. Besides, this OB corresponds to two 12 dB (16*16 subscribers) power splitters, taking into account mux/demux, propagation and circulator losses. A compromise between split ratio and range needs to be considered. Thus, one of the two 12 dB budget increase can also allow reach extension between the CO and the remote node (including the 25 km reach extension). However, propagation effects such as RBS and dispersion in the fibre would limit this extension. A reduction in RBS level is also needed to improve the performance of this configuration. Different solutions have been studied to reduce the RBS level such as: frequency modulation of the laser source or applying bias dithering at the RSOA.

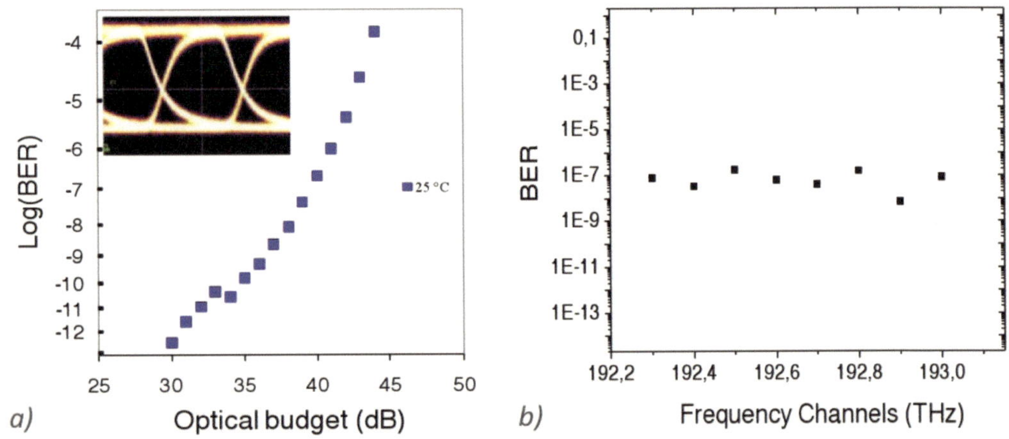

Fig. 15. (a) BER as a function of the optical budget. Inset: 2.5 Gbit/s eye-diagram at the output of the RSOA driven at 90 mA and with an input power of -10 dBm (b) BER values for different λ-channels for an optical budget of 40 dB, or a Rx input power of -30 dBm (de Valicourt et al., 2010a)

A cost effective hybrid WDM/TDM-PON which can potentially feed 2048 subscribers (16×16×8 = 2048 subscribers) at a data rate of 2.5 Gbit/s is presented in this section. The large gain and high output power of the RSOA have also allowed extending the link reach up to 45 km instead of the standard 20 km. However, these achievements are obtained at the expense of an increase in deployment and operation costs. We believe this solution is economically viable since these costs are shared between many users, and multi-wavelength sources are becoming cheaper with the advent of Photonic Integrated Circuits (PIC). This 2.5 Gbit/s upstream colourless result allows investigating this solution to achieve in the trunk line a wavelength multiplex of several next generation access solutions (10 Gbit/s down- and 2.5 Gbit/s up-stream).

6. Limitations and improvements

Architecture based on single-fibre bidirectional link seems the most interesting and cost efficient approach. ONU becomes a key element for the network evolution. Transparent and flexible architecture based on WDM technology is necessary thus colourless ONU need to be available. High gain should be provided by the transmitter to reach the necessary optical budget and high modulation speed is needed. Bit rates up to 10 Gbit/s (per wavelength) are required to follow the evolution of the 10 Gbit/s GPON. RSOA could be the missing building block to reach this ideal network. However the modulation bandwidth and the chirp are still issues that need to be solved.

6.1 Long reach PON using low chirp RSOA

Another question about Hybrid PON is its property to be compatible with long reach network configuration. It was shown in the previous section that high gain RSOAs enable high optical budget, for instance, up to 36 dB and 45km transmission at 2.5 Gbit/s. A high optical budget is necessary to obtain a long reach PON (compensation of the fibre attenuation). The limitation imposed on the bit rate and distance by the fibre dispersion can dramatically increase depending on the spectral width of the source. This problem can be overcome by reducing the chirp produced by the RSOA device. Chirp reduction was demonstrated using a 2-section RSOA and how it can be used to reduce the transmission penalties (de Valicourt et al., 2010b). We propose an extended reach hybrid PON, taking advantages of a very high gain Reflective Semiconductor Optical Amplifier (RSOA) and the two-electrode configuration operating at 2.5 Gbit/s (de Valicourt et al., 2010c).

Two RSOAs with a cavity length of 500 µm are used in the experimental setup, one with mono and the other with dual-electrode configuration. The dual-electrode RSOA was realized by proton implantation in order to separate both electrodes. The single electrode RSOA was driven at 60 mA and the dual-electrode at 20 mA on the input electrode and 115 mA at the mirror electrode. Both current values correspond to optimized conditions in order to obtain low transmission penalties. It is to be noted that in a dual-electrode RSOA, the AC current is applied to the input/output electrode. In both cases, the CW optical input power was -8.5 dBm. Fig. 16 displays BER measurements performed at 1554 nm and 2.5 Gbps as a function of the received power for one electrode and two-electrode RSOA. The penalties due to 100 km transmission with a single electrode RSOA do not enable to reach the FEC limit. From 25 km to 50 km (100 km), we obtain penalties of 1.2 (3.4) dB. One can see that a BER of 10^{-4} (FEC limit) has thus been measured with bi-electrode RSOA at a received optical power of -35 dBm over 100 km SMF. The penalties due to extended 25 and 50 km SMF are much lower than with single electrode RSOA (respectively 0.5 and 1.4 dB). These transmission

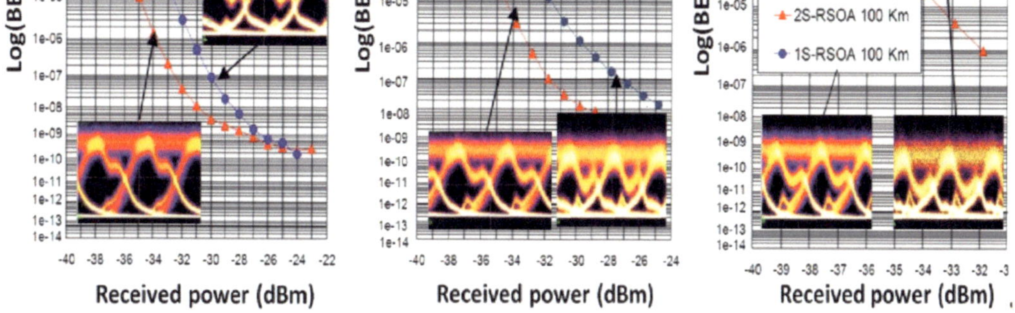

Fig. 16. Comparison of BER value as a function of the received power for mono-electrode and bi-electrode RSOA over 50, 75 and 100 km at 2.5 Gbps (de Valicourt et al., 2010c).

results clearly show the correlation between the penalty and the chirp. The latter has more pronounced effects over long SMF. In Figure 16, a comparison between single and bi-electrode RSOA over 100 km transmission is shown and a difference of 4.1 dB is obtained at the FEC limit. We can clearly see that the eye diagram starts to be closed due to the chirp on single-electrode devices over long distances. This effect is much reduced when using bi-electrode RSOAs which confirms BER measurements. The proposed network design allows the use of Dense-WDM (DWDM) which means 62 wavelengths considering the 50 nm optical bandwidth of the RSOA. By considering the passive splitter (four clients), 248 potential suscribers can be feeded. At the FEC level, a variable attenuation of 4 dB is obtained which can be use as a splitter in order to design two parallels WDM PONs (2*248=496 customers) over 100Km.

It was shown that the large gain of the RSOA and also the low chirp allows a reach extension of the link from standard 20km to 100 km. We demonstrated that penalties due to the transmission over 100 km SMF at 2.5 Gbit/s are reduced using an optimized multi-electrode device and a BER below the FEC limit was achieved. We also believe this effect will be even more pronounced when 10 Gbit/s RSOA will be used.

6.2 Reaching 10 Gbit/s modulation without any electronic processing

Active research on high bit rate RSOA has led to 10 Gbit/s operation with EDC (Torrientes et al., 2010), OFDM (Duong et al., 2008) or electronic filtering (Schrenk et al, 2010). Bandwidth improvement to 7 GHz small-signal bandwidth with dual-electrode devices have been obtained but no large signal operation (Brenot et al, 2007). However the modulation bandwidth of one-section RSOA is limited to 2 GHz and increasing the modulation bandwidth of RSOA is still a challenge. Since carrier lifetime is mainly governed by stimulated emission rate, we have decided to increase the length of our RSOA to increase photon density, hence reducing carrier lifetime (de Valicourt et al., 2011). This device was chosen because an open eye diagram was obtained when the RSOA was driven by a 2^7-1 PRBS at 10 Gbit/s (Figure 17. (a)). The experimental set-up used for the 10 Gbit/s modulation is the same as represented in Figure 13. An ECL is used to launch a 4.5 dBm CW signal into the system through an optical circulator (OC). The signal is coupled into the RSOA which is modulated and generates the upstream signal. The RSOA is driven by a 2^7-1 PRBS at 10 Gbit/s, with a DC bias of 160 mA. The upstream signal propagates on various SMF lengths. A variable optical attenuator is placed in front of the receiver in order to analyze the performance of the system as a function of the received power. BER measurements are done using an APD receiver and an error analyzer. BER measurements without ECL have led to a BER floor of 10^{-6} (ASE regime).

With optical injection, BER values below the FEC limit in BtB and after 2km transmission are obtained (Figure 17. (b)). Error-free operation can either be obtained with FEC codes, or under certain optical injection regimes. However we can clearly see that the eye diagram tends to be closed due to the chirp over long distances. Multi-electrode devices can be used in order to compensate for this effect as demonstrated in the previous section.

As described in section 4, the modulation speed of RSOA is limited by the carrier lifetime. In the large signal regime, the slow decay is probably governed by the no-stimulated recombination process, which increases the carrier lifetime. A 3 GHz modulation bandwidth can be obtained with 850 μm long RSOA, which has led to the first eye-opening of a RSOA at 10 Gbit/s without electrical equalization or passive electronic filtering. Limitation due to the chirp is observed and further works are underway to overcome this effect using multi-electrodes devices. Longer devices and dual-electrode devices will be studied to improve the modulation and transmission properties.

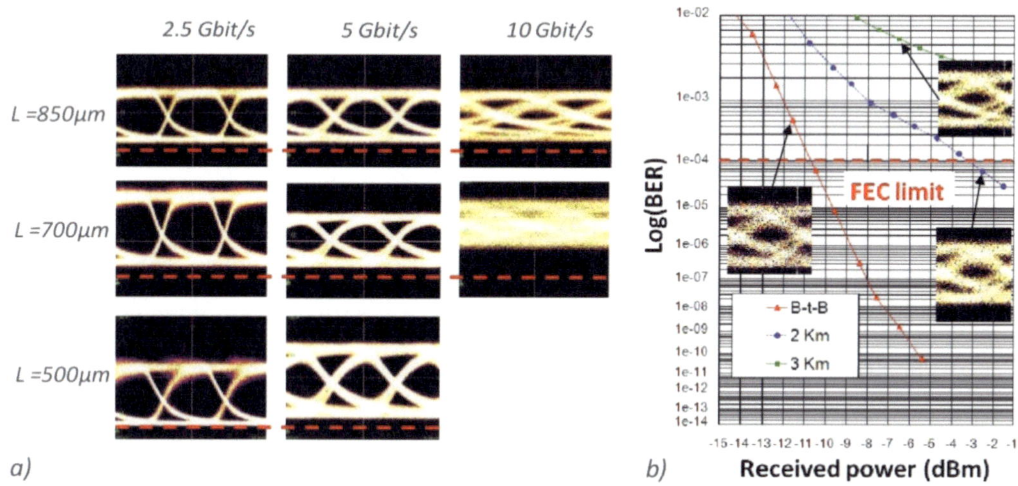

Fig. 17. (a) Eye diagrams at various bit rates of RSOA whose length varies from 500 to 850 μm. The collected power is pure ASE. Red lines are the dark levels. (b) BER value as a function of the received power for 850 μm long RSOA modulated at 10 Gbit/s (de Valicourt et al., 2011)

7. Conclusion

Nowadays, research in the telecom area is partly focused on passive optical network architecture. WDM-PON seems to be a promising approach allowing high data bit rate and flexibility. WDM techniques used in long-haul systems are now mature, however the shift to local access networks is more challenging. New requirements appear such as cost reduction, the need for new key devices at the ONU and compatibility with the existing network. Colourless ONU are necessary to obtain cost effective architecture and RSOA is one potential solution. In this chapter, we focused on these devices.

The SOA theory has been discussed and applied to Reflective SOA devices. We underline several physical mechanisms that are responsible for the carrier density variation. The stimulated, radiative and non-radiative recombination rates are described. A model has been developed, taking into account several longitudinal sub-sections of the active guide. RSOAs exhibit a non-homogeneous carrier density profile which strongly affects the overall gain. At the input/output of the device, a strong saturation effect is observed. Therefore the net gain needs to be carefully integrated along the device taking in account this non-homogeneity. All these results confirm the presence of key parameters such as the length and the optical confinement which should lead to design rules.

To assess the RSOA dynamics, the carrier lifetime is estimated. The E/O modulation bandwidth mainly depends on this parameter, for instance shorter carrier lifetime induces larger 3 dB E/O modulation bandwidth. The reduction of the carrier lifetime is required to obtain high speed RSOAs.

Potential cost effective solutions for next generation of access network could be based on RSOA devices. Therefore, research on RSOA devices is driven by WDM-PON applications. It is of prime interest to solve issues related to this application. RSOAs as colourless ONU have been investigated for access network. High performances RSOAs enable an upstream transmission of 8 WDM channel at 2.5 Gbit/s over 45 km. A high optical budget (36 dB) was demonstrated.

The chirp remains one major limiting factor as well as the modulation speed. 2-section RSOAs were used to overcome the first drawback. The use of 2-section RSOAs allows a 100 km transmission below the FEC limit at 2.5 Gbit/s. Finally, long RSOA allows performing the first direct 10 Gbit/s modulation with open eye diagram thanks to the E/O modulation bandwidth increase.

Therefore, as a general conclusion, RSOAs show a great potential as a next generation of optical transmitter. It is a colorless device which can be used in WDM access networks. However the modulation speed is still limited and 10 Gbit/s modulation needs to be realised over a minimum of ten kilometres.

Tech-eco analysis has to be performed in order to evaluate the different technologies for WDM-PON and a trade-off between performances and cost will determine the future of optical access network. RSOA are still limited in terms of performances and architecture but new approach such as self-seeding could overcome these main issues.

8. Acknowledgements

The work in this chapter would not have been possible without the support of numerous people and I would like to acknowledge a few of them here. Firstly, the author would like to thank Dr. Romain Brenot for his guidance and advices.

Next I would like to thank fellow workers at III-V lab, especially Francis Poingt and Marco Lamponi who worked closely with me. This collaboration was key to the success of this study. Additionally, I wish to acknowledge Dr. Philippe Chanclou for the fruitful discussions.

9. References

An F.T, Soo Kim K., Gutierrez D., Yam S., Hu E., Shrikhande K., & Kazovsky L.G. (2004),"SUCCESS: A Next-Generation Hybrid WDM/TDM Optical Access Network Architecture", *J. Lightwave Technol.*, Vol. 22, No. 11, November 2004

Brenot R., Pommereau F., Le Gouezigou O., Landreau J., Poingt F., Le Gouezigou L., Rousseau B., Lelarge F., Martin F., & Duan G-H (2005), "Experimental study of the impact of optical confinement on saturation effects in SOA", *in proc. OFC 2005*, OME50, March 2005

Brenot R., Provost J.-G., Legouezigou O., Landreau J., Pommereau F., Poingt F., Legouezigou L., Derouin E., Drisse O., Rousseau B., Martin F., Lelarge F., & Duan G.H. (2007), "High modulation bandwidth reflective SOA for optical access networks", *in Proc. ECOC*, 2007, pp. 1-2, Berlin, Germany

Buldawoo N., Mottet S., Le Gall F., Sigonge D., Meichenin D., & Chelles S. (1997), "A Semiconductor Laser Amplifier-Reflector for the future FTTH Applications", *in Proc. ECOC'97*, Sept. 1997, pp. 196-199, Edinburgh, UK

Chanclou P., Payoux F., Soret T., Genay N., Brenot R., Blache F., Goix M., Landreau J., Legouezigou O., & Mallécot F. (2007), "Demonstration of RSOA-based remote modulation at 2.5 and 5 Gbit/s for WDM PON", *in Proc. OFC*, OWD1, 2007, Anaheim, USA

Connelly M. J., "Wide-Band Steady-State Numerical Model and Parameter Extraction of a Tensile-Strained Bulk Semiconductor Optical Amplifier," *IEEE J. Quantum Electroni.*, Vol. 43, No. 1, Jan. 2007, pp. 47-56

D'Alessandro D., Giuliani G., & Donati S. (2001), "Spectral gain and noise evaluation of SOA and SOA-based switch matrix", *IEE Proc.-Optoelectron.*, Vol. 148, No. 3, June 2011, pp. 125-130

de Valicourt G., Maké D., Fortin C., Enard A., Van Dijk F., & Brenot R. (2011), "10Gbit/s modulation of Reflective SOA without any electronic processing", *in Proc. OFC'11*, OThT2, 2011, Los Angeles, USA

de Valicourt G., Make D., Lamponi M., Duan G., Chanclou P. , & Brenot R. (2010a), "High Gain (30 dB) and High Saturation Power (11dBm) RSOA Devices as Colourless ONU Sources in Long Reach Hybrid WDM/TDM -PON Architecture", *Photonics technology letters*, Vol. 22, No. 3, Feb. 2010, pp. 191-193

de Valicourt G., Pommereau F., Poingt F., Lamponi M., Duan G.H., Chanclou P., Violas M. A., & Brenot R. (2010b) "Chirp Reduction in Directly Modulated Multielectrode RSOA Devices in Passive Optical Networks", *Photonics technology letters*, Vol. 22, No. 19, Oct. 1, 2010, pp. 1425-1427

de Valicourt G., Lamponi M., Duan G.H., Poingt F., Faugeron M., Chanclou P., & Brenot R. (2010c), " First 100 km uplink transmission at 2.5 Gbit/s for hybrid WDM/TDM PON based on optimized bi-electrode RSOA", *in Proc. ECOC'10*, Tu.5.B.6, 2010, Torino, Italy

de Valicourt G., Maké D., Landreau J., Lamponi M., Duan G.H., Chanclou P. , & Brenot R. (2009), "New RSOA Devices for Extended Reach and High Capacity Hybrid TDM/WDM -PON Networks", *in Proc. ECOC'09*, P9.5.2, 2009, Vienna, Austria

Duong T., Genay N., Chanclou P., Charbonnier B., Pizzinat A., & Brenot R., "Experimental demonstration of 10 Gbit/s upstream transmission by remote modulation of 1 GHz RSOA using adaptively modulated optical OFDM for WDM-PON single fiber architecture," *in Proc. ECOC*, Th.3.F, Sep. 2008, Brussels, Belgium

Feuer M., Wiesenfeld J., Perino J., Burrus C., Raybon G., Shunk S., & Dutta N. (1996), "Single-port laser-amplifier modulators for local access", *Photonics technology letters*, Vol.8, No.9, Sept. 1996, pp.1175-1177

Healey P., Townsend P., Ford C., Johnston L., Townley P., Lealman I., Rivers L., Perrin S. , & Moore R. (2001), "Spectral slicing WDM-PON using wavelength-seeded reflective SOAs", *Electron. Lett.*, Vol. 37, No. 19, Sept. 2001, pp. 1181 - 1182

Iezekiel S., Elamaran B., & Pollard R.D. (2000), "Recent developments in lightwave network analysis", *Engineering Science and Education Journal*, Vol. 09, No. 06, Dec. 2000, pp. 247-257

Kao K. C. & Hockham G. A. (1996), Dielectric-fibre surface waveguides for optical frequencies, *Proc. IEE*, Vol. 113, No. 7, July 1966, pp. 1151-1158

Koponen J.J., & Söderlund M.J. (2004), "A duplex WDM passive optical network with 1:16 power split using reflective SOA remodulator at ONU", *in proc. OFC'04*, MF 99, March 2004, Los Angeles, USA

Lee H-H., Cho S-H., Lee J-H., Jung E-S., Yu J-H., Kim B-W., Lee S-S., Lee S-H., Koh B-H., Sung J-S., Kang S-J., Kim J-H., & Jeong K-T. (2009), "First Commercial Service of a Colorless Gigabit WDM/TDM Hybrid PON System", *in Proc. OFC*, PDPD9, 2009, San Diego, USA

Lee W. R., Park M. Y., Cho S. H., Lee J. H., Kim C. Y., Jeong G., & Kim B. W. (2005), "Bidirectional WDM-PON based on gain-saturated reflective semiconductor optical amplifiers", *Photonics technology letters*, Vol. 17, No. 11, Nov. 2005, pp. 2460–2462

Lelarge F., Chimot N., Rousseau B., Martin F., Brenot R., & Accard A. (2010), "Chirp Optimization Of 1550 nm InAs/InP Quantum Dash Based Directly Modulated Lasers For 10Gb/s SMF Transmission Up To 65Km" *in proc. IPRM*, May-June, 2010, pp. 1-3, Kagawa, Japan

Liu Z., Sadeghi M., de Valicourt G., Brenot R., & Violas M.(2011), "Experimental Validation of a Reflective Semiconductor Optical Amplifier Model used as a Modulator in Radio over Fiber Systems", *Photonics technology letters*, Vol. 23, No. 9, May 1, 2011, pp. 576-578

Marazzi L., Parolari P., de Valicourt G., & Martinelli M. (2011), "Network-Embedded Self-Tuning Cavity for WDM-PON Transmitter", *in proc. ECOC*, Mo.2.C.3, Sept. 2011, Geneva, Switzerland

Oh J., Koo S. G., Lee D., & Park S-J. (2007), "Enhanced System Performance of an RSOA Based Hybrid WDM/TDM-PON System Using a remotely Pumped Erbium-Doped Fiber Amplifier", *in Proc. OFC*, PDP9, 2007, Anaheim, USA

Olshansky R., Su C. A., Manning J., & Powazinik W. (1984), "Measurement of radiative and nonradiative recombination rates in InGaAsP and AlGaAs light sources", *J. Quantum Electron.*, Vol. 20, No. 8, 1984, pp. 838-854

Olsson N. A. (1988), "Polarisation-independent configuration optical amplifier", *Electron. Lett.*, Vol. 24, No. 17, Aug. 1988, pp. 1075-1076

Omella M., Polo V., Lazaro J., Schrenk B., & Prat J. (2008), "10 Gb/s RSOA Transmission by Direct Duobinary Modulation", *in Proc. ECOC'08*, Tu.3.E.4, Sept. 2008, Brussels, Belgium

Otani T., Goto K., Abe H., Tanaka M., Yamamoto H., & Wakabayashi H. (1995), "5.3 Gbit/s 11300 km data transmission using actual submarine cables and repeaters", *Electron. Lett.*, Vol. 31, No. 5, 1995, pp. 380-381

Prat J., Arellano C., Polo V., & Bock C. (2005), "Optical network unit based on a bidirectional reflective semiconductor optical amplifier for fiber-to-the-home networks", *Photonics technology letters*, Vol.17, No.1, January 2005, pp.250-252

Schrenk B., de Valicourt G., Omella M., Lazaro J., Brenot R., & Prat J. (2010), "Direct 10 Gb/s Modulation of a Single-Section RSOA in PONs with High Optical Budget", *Photonics technology letters*, Vol. 22, No. 6, March 15, 2010, pp. 392-394

Tanaka S., Tomabechi S., Uetake A., Ekama M., & Morito K. (2006), Recard high saturation output power (+20 dBm) and Low NF (6.0 dB) polarization-insensitive MQW-SOA module, *IET Electronics Letters*, Vol. 42, No. 18, August 2006, pp. 1050-1060

Torrientes D., Chanclou, P., Laurent F., Tsyier S., Chang Y., Charbonnier B., & Raharimanitra F. (2010), RSOA-Based 10.3 Gbit/s WDM-PON with Pre-Amplification and Electronic Equalization, *in proc. OFC*, JThA28, 2010, San Diego, USA

Trung Le Q., Deniel Q., Saliou F., de Valicourt G., Brenot R., & Chanclou P. (2011), RSOA-based External Cavity Laser as Cost-effective Upstream Transmitter for WDM Passive Optical Network, *CLEO 2011*, JWA9, May 1-6, 2011, Baltimore, USA

Wong, E., Ka Lun Lee, & Anderson T.B. (2007), Directly Modulated Self-Seeding Reflective Semiconductor Optical Amplifiers as Colorless Transmitters in Wavelength Division Multiplexed Passive Optical Networks, *J. Lightwave Technol.*, Vol. 25, No. 1, January 2007

2

High-Speed All-Optical Switches Based on Cascaded SOAs

Xuelin Yang, Qiwei Weng and Weisheng Hu
The State Key Laboratory of Advanced Optical Communication Systems and Networks, Shanghai Jiao Tong University, Shanghai, China

1. Introduction

Lots of research efforts have been focused to realize all-optical high-speed switches through nonlinear optical elements, for instance, high nonlinear fibers (HNLF), nonlinear waveguides as well as semiconductor optical amplifiers (SOAs). All-optical switches incorporating SOAs is one of the particularly attractive candidates due to their small size, high nonlinearities (low switching energy required) and ease of integration. All-optical switches also keep the network transparent, enhance the flexibility and capacity in network, and offer the function of signal regeneration, therefore SOAs provide various attractive all-optical functions in high-speed signal processing in fiber communication systems (Stubkjaer, 2000; Poustie, 2007), including all-optical AND/XOR logic gates, wavelength conversion (WC), optical-time division multiplexing (OTDM) de-multiplexing, optical signal regeneration and so on, which will be essential to the implementation of future wavelength division multiplexing (WDM) or optical packet switching (OPS) networks.
However, the operation speed of SOA based switches is inherently limited by its relative slow carrier lifetime (in an order of 100 ps) (Manning et al., 2007). Various schemes have been proposed to enhance the operation speed of SOA-based all-optical devices, for instance, 160 Gb/s and 320 Gb/s wavelength conversion was reported by using a detuned narrow band-pass filter to spectrally select one of the side-bands (blue-shifted or red-shifted) of the output signal (Liu et al., 2006, 2007). In this case, the SOA operation speed can be increased via the chirp effect on the SOA output associated with the SOA ultrafast gain dynamics. It has been shown that, the CW modulation response time has been reduced from 100 ps to 6 ps via filter detuning (Liu et al., 2006, 2007). Although using a detuned filter after the SOA can improve the optical signal-to-noise ratio (OSNR) of the output when comparing with the case of using a non-detuned filter (Leuthold, 2002), however the OSNR of the output signal will degrade to a large extent since the optical carrier was suppressed.
Recently, all-optical high-speed switches based on the cascaded SOAs were proposed and demonstrated. In Fig. 1, an all-optical switch incorporating two cascaded SOAs was proposed as an alternative high-speed technique, which was dubbed as "turbo-switch" (Manning et al., 2006; Yang et al., 2006, 2010), while preserving the OSNR of the output signal. An error-free wavelength conversion was demonstrated at 170 Gb/s (Manning et al., 2006). In addition, the operating speed of an all-optical XOR gate was also demonstrated at

85Gb/s, where dual ultrafast nonlinear interferometers (UNIs) were implemented (Yang et al., 2006, 2010) and the turbo-switch configuration was incorporated.

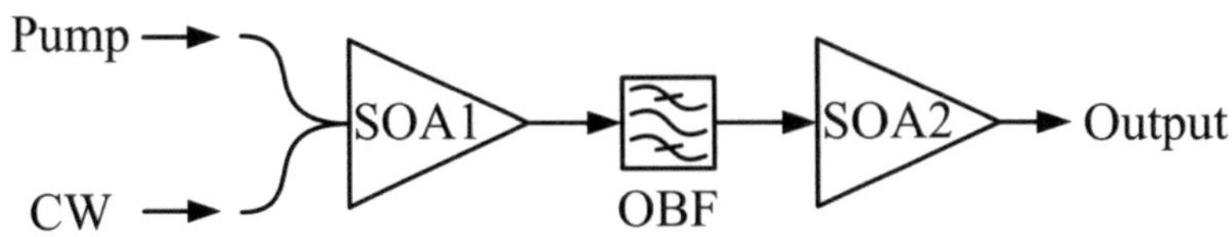

Fig. 1. Schematic setup of the turbo-switch, where the OBF is used to remove the pump signal. OBF: optical band-pass filter.

In this chapter, we will review the recent progress of the all-optical high-speed switches using cascaded SOAs, from both theoretical and experimental aspects. A majority of the publications (Manning et al., 2006, 2007; Yang et al., 2006, 2010) related to turbo-switch were reported, showing the high-speed experimental performances of turbo-switch over a single SOA. Apparently, a systematic theoretical turbo-switch model is necessary for the purpose of understanding the fundamental behaviors of the turbo-switch and how to further enhance the switch performance. First of all, we will present a detailed time-domain SOA model, from which the turbo-switches and switches with three or more cascaded SOAs can be evaluated. For the reason of convenience, we will refer hereafter to this kind of switch, including turbo-switch, as cascaded-SOA-switch. Then, we will focus on the relation between the overall performance of the switch and the nonlinear gain/refractive-index dynamics of the individual SOAs. The amplitude/phase dynamics of the optical output signal from the switch will be analyzed in details and compared with the experimental data. The SOA model will certainly help us not only to understand the basic principles of the switch, but also to exploit the way and the critical conditions for the switch to operate at even higher bit-rates.

The chapter is organized as follows. Section 2 presents a comprehensive theoretical analysis of the cascaded-SOA-switch, where the SOA model and the corresponding simulation method are presented. Simulation results including the gain/phase dynamics, pattern effect mitigation using turbo-switch, are shown in Section 3. Experimental demonstrations of 170 Gb/s AND gate (wavelength conversion) and 85 Gb/s XOR gate using turbo-switches are presented in Section 4. The cascaded-SOA-switches are further exploited in terms of the number of cascaded SOAs in Section 5, where the overall gain recovery time, the noise figure as well as the impact of injected SOA current of the cascaded switches are illustrated in details, as simulated by the model. Finally, conclusions will be given in Section 6.

2. Theoretical analysis of SOAs

To explore the operation principle and understand the performances of the cascaded-SOA-switch, a time-domain SOA model is required to analyze the fundamental gain/phase behaviors of the SOA-based device as well as to simulate the speed and application of the devices.

2.1 SOA model

The basic time-domain rate equations describing the carrier dynamics via the inter-band and intra-band processes in a single SOA, as proposed in (Gutiérrez-Castrejón, 2009; Mecozzi &

Mørk, 1997), are adopted. Travelling-wave equations in terms of the optical amplitude/power and phase, derived from Maxwell equations and Kramers-Kronig relations, are also incorporated in the SOA model to obtain the amplitude and phase of the output optical signal propagating through the SOA (Mecozzi & Mørk, 1997; Agrawal & Olsson, 1989).

Following the SOA model in (Mecozzi & Mørk, 1997), rate equations for the total carrier density N related to the (inter-band) band-filling effect, and the local carrier density variations n_{CH} and n_{SHB}, which are associated with the ultrafast (intra-band) effects: carrier heating (CH) and spectrum hole burning (SHB) processes respectively, can be expressed as follows:

$$\frac{\partial N(z,t)}{\partial t} = \frac{I}{eV} - R(N(z,t)) - v_g g S(z,t) - v_g g_{ase}\left[S_{ase}^{+}(z,t) + S_{ase}^{-}(z,t)\right] \tag{1}$$

$$\frac{\partial n_{CH}(z,t)}{\partial t} = -\frac{n_{CH}(z,t)}{\tau_{CH}} - \frac{\varepsilon_{CH}}{a_0 \tau_{CH}} g S(z,t) \tag{2}$$

$$\frac{\partial n_{SHB}(z,t)}{\partial t} = -\frac{n_{SHB}(z,t)}{\tau_{SHB}} - \frac{\varepsilon_{SHB}}{a_0 \tau_{SHB}} g S(z,t) - \left[\frac{\partial N(z,t)}{\partial t} + \frac{\partial n_{CH}(z,t)}{\partial t}\right] \tag{3}$$

where the first term in the right hand side (RHS) of (1) represents the increase of the total carrier density due to the injected current I to the SOA. Here, we have assumed a uniform distribution of the injected current along the longitude. In (1), e is the electron charge, and V is the volume of the active region in the SOA.

The radiative and nonradiative recombination rate due to the limited carrier lifetime in the SOA, $R(N)$ (Connelly, 2001), can be approached by,

$$R(N) = AN + BN^2 + CN^3 \tag{4}$$

where A, B, C represent the linear, bimolecular, and auger recombination coefficients respectively.

The third and fourth terms in the RHS of (1) are used to account for the depletion of total carrier density aroused from the stimulation emission by the injected light and the amplified spontaneous emission (ASE), respectively. v_g is the group velocity. g is the gain coefficient and S is the photon density in the active region. g_{ase} is the equivalent gain coefficient for ASE (Talli & Adams, 2003). τ_{CH} and ε_{CH} in (2) are carrier-carrier relaxation time and gain suppression factor caused by CH, while τ_{SHB} and ε_{SHB} in (3) are temperature relaxation time and gain suppression factor caused by SHB.

To take the gain dispersion into account better, and make our model applicable in a wide optical wavelength range, a polynomial model for the gain coefficient (Leuthold et al., 2000), which combines of a quadratic and a cubic function, is used, with one modification to include the ultrafast effect induced by CH and SHB.

$$g = \begin{cases} g_l + g_h, & \lambda < \lambda_z(N) \\ 0, & \lambda \geq \lambda_z(N) \end{cases} \tag{5a}$$

$$g_\beta = c_{N,\beta}\left[\lambda - \lambda_z(N)\right]^2 + d_{N,\beta}\left[\lambda - \lambda_z(N)\right]^3 \tag{5b}$$

where β = l, h represents the gain coefficient attributed to total carrier density N and CH/SHB effect, respectively.
Polynomial coefficients are calculated by,

$$c_{N,\beta} = 3\frac{g_{p,\beta}}{\left[\lambda_z(N) - \lambda_p(N)\right]^2} \tag{5c}$$

$$d_{N,\beta} = 2\frac{g_{p,\beta}}{\left[\lambda_z(N) - \lambda_p(N)\right]^3} \tag{5d}$$

where g_p, β, $\lambda_{p(N)}$ and $\lambda_{z(N)}$ stand for the material gain at the peak wavelength, the shifted wavelength at peak and transparency respectively. They are approximated by,

$$g_{p,l} = a_0(N - N_0) + \bar{a}a_0N_0e^{-N/N_0} \tag{5e}$$

$$g_{p,h} = a_0(n_{CH} + n_{SHB}) \tag{5f}$$

$$\lambda_p(N) = \lambda_{p_0} - \left[b_0(N - N_0) + b_1(N - N_0)^2\right] \tag{5g}$$

$$\lambda_z(N) = \lambda_{z_0} - z_0(N - N_0) \tag{5h}$$

where a_0, N_0, $\bar{a}$, λ_{p0}, b_0, b_1, λ_{z0}, and z_0 are parameters which have to be obtained by experimental gain dispersion curves (Leuthold et al., 2000). N_0 represents the transparency carrier density at the peak wavelength λ_0.
By definition, the photon density S (in unit of m^{-3}) in (1)-(3) can be expressed in terms of the light power P (in unit of W) as,

$$S(z,t) = \frac{P(z,t)}{h(c/\lambda)(\delta/\Gamma)v_g} \tag{6}$$

where h, c, δ, Γ denotes for Planck's constant, speed of light in vacuum, cross section area of the active region and confinement factor, respectively.
The travelling-wave equation of the input optical light (Agrawal & Olsson, 1989) is,

$$\frac{\partial P(z,t)}{\partial z} + \frac{1}{v_g}\frac{\partial P(z,t)}{\partial t} = (\Gamma g - \alpha_{\text{int}})P(z,t) \tag{7}$$

where the power P is a function of time t and position z along the active waveguide (z-axis) of the SOA. α_{int} is the internal loss in the active region. Eq. (7) only represents the positive direction propagation of the input light, since the facet reflection of the SOA (below 10^{-4}) is usually ignorable (Dutta & Wang, 2006).
For the propagation of the ASE power inside the amplifier, a bi-directional model presented in (Talli & Adams, 2003) is adopted, where the ASE is described by its total power while neglecting its spectral dependency. Equivalent coupling efficiency β_{ase}, equivalent wavelength λ_{ase}, and equivalent gain coefficient g_{ase}, are used in the calculation, for the reason of computational efficiency.

$$\frac{\partial P_{ase}^{\pm}(z,t)}{\partial z} \pm \frac{1}{v_g}\frac{\partial P_{ase}^{\pm}(z,t)}{\partial t} = \pm(\Gamma g_{ase} - \alpha_{\mathrm{int}})P_{ase}^{\pm}(z,t) \pm \beta_{ase}R_{sp}\frac{hc}{\lambda_{ase}}\frac{\delta}{\Gamma} \tag{8}$$

where an additional term in the RHS, comparing to (7), is used to account for the spontaneous emission (SE) coupled into the effective waveguide. $R_{sp} = BN^2$ is the SE rate. "+" stands for the co-propagating direction with the input light, while "-" represents the counter-propagating direction.

Carrier density variations not only affect the gain, but also change the phase of the input optical signal. Associated with the gain dynamics through Kramers-Kronig relations, the phase shift (Mecozzi & Mørk, 1997) of the optical beam due to the SOA nonlinearity can be expressed as,

$$\frac{\partial \phi(z,t)}{\partial z} + \frac{1}{v_g}\frac{\partial \phi(z,t)}{\partial t} = -\frac{1}{2}\Gamma\left[\alpha_N g_l + \alpha_T g_{h,n_{SHB}=0}\right] \tag{9}$$

where α_N and α_T is the α-factors (also known as linewidth enhancement factor) for the band-filling and CH process, respectively. The subscript $n_{SHB} = 0$ means the SHB impact on the phase shift is ignored here.

It should be mentioned that, many physical effects of the SOA, including two-photon absorption (TPA), ultrafast nonlinear refraction (UNR), free-carrier absorption (FCA) and group velocity dispersion (GVD), are neglected in our SOA model. Ultrafast processes such as TPA, FCA and UNR are ignored reasonably, because these effects become important only when pulse energy is stronger than 1 pJ (Yang et al., 2003), while the pulse energy used in our simulation is generally lower than 0.1 pJ. GVD is also neglected, since the Gaussian pump pulsewidth (full width at half maximum, FWHM) in the paper is assumed to be 2~3 ps, which means that the spectral detuning from the central frequency is less than a few THz (Mecozzi & Mørk, 1997).

2.2 Numerical method

In order to solve the model numerically, we divide the SOA into N_z sections of equal length in the optical active waveguide, thus having a section length of $\Delta z = L/N_z$, and choose a corresponding time resolution of $\Delta t = \Delta z/n_g$. N_z should be large enough to have a good numerical approximation.

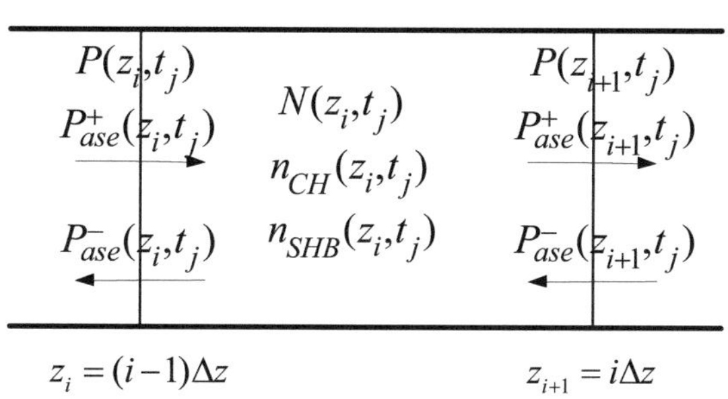

Fig. 2. A schematic sketch of the ith section of the SOA.

Fig. 2 shows a sketch of the ith section in the SOA, where i=1,2,…, N_z and j=1,2,…,N_t. N_z and N_t are the total number of the SOA sections and time steps respectively (Connelly, 2001).

Optical powers and ASE propagating in the positive and negative directions are calculated at the boundaries of each section, while the total carrier density and local carrier changes caused by the CH and SHB processes are considered at the center of each section. When the time interval Δt is small enough, the left hand side (LHS) of (1) can be approximated by,

$$\frac{\partial N(z_i,t_j)}{\partial t}=\frac{N(z_i,t_j)-N(z_i,t_{j-1})}{\Delta t} \tag{10}$$

Thus, basing upon the carrier density and the photon densities at the previous time step, we have,

$$\begin{aligned} N(z_i,t_j)=N(z_i,t_{j-1})+\Delta t\Bigg[&\frac{I}{eV}+R(N(z_i,t_{j-1}))-v_g g\frac{S(z_i,t_{j-1})+S(z_{i+1},t_{j-1})}{2}\\ &-v_g g_{ase}\frac{S^+_{ase}(z_i,t_{j-1})+S^+_{ase}(z_{i+1},t_{j-1})}{2}-v_g g_{ase}\frac{S^-_{ase}(z_i,t_{j-1})+S^-_{ase}(z_{i+1},t_{j-1})}{2}\Bigg] \end{aligned} \tag{11}$$

where a linear interpolation is employed to estimate the photon densities of the input optical beam, co-propagating and counter-propagating ASEs at the center of each section. Similar method can be applied to (2) and (3), to calculate the local carrier density variations due to CH and SHB processes.

The first term in the LHS of (7), describes the optical power propagating along the z-axis of the SOA, and experiencing an exponential amplification by a factor of (Γg - α_{int}), as shown in the RHS, which can be assumed constant in a sufficiently small interval Δz. The second term in the LHS, however, accounts for the optical power variation during the travelling time period in the section, which can be included using values obtained at last time step (Bischoff, 2004). Therefore, a solution of (7) is,

$$P(z_{i+1},t_j)=P(z_i,t_{j-1})\exp\left\{\left(\Gamma g N(z_i,t_{j-1})-\alpha_{\text{int}}\right)\Delta z\right\} \tag{12}$$

subjected to boundary condition,

$$P(z_1,t_j)=P_{in}(t_j) \tag{13}$$

where $P_{in}(t_j)$ denotes the input optical power at t_j.

Similar solutions can be given for the co-propagating and the counter-propagating ASEs, as described in (8),

$$P^+_{ase}(z_{i+1},t_j)=P^+_{ase}(z_i,t_{j-1})\exp\{g'_{ase}\Delta z\}+\left[\beta_{ase}R_{sp}(N(z_i,t_{j-1}))\frac{hc}{\lambda_{ase}}\frac{\delta}{\Gamma}\right]\frac{\exp\{g'_{ase}\Delta z\}-1}{g'_{ase}} \tag{14a}$$

$$P^-_{ase}(z_i,t_j)=P^-_{ase}(z_{i+1},t_{j-1})\exp\{g'_{ase}\Delta z\}+\left[\beta_{ase}R_{sp}(N(z_i,t_{j-1}))\frac{hc}{\lambda_{ase}}\frac{\delta}{\Gamma}\right]\frac{\exp\{g'_{ase}\Delta z\}-1}{g'_{ase}} \tag{14b}$$

where $g'_{ase}=\Gamma g(N(z_i,t_{j-1}),\lambda_{ase})-\alpha_{\text{int}}$, and subjected to boundary conditions respectively,

$$P^+_{ase}(z_1,t_j)=0 \tag{15a}$$

$$P_{ase}^{-}(z_{Nz+1}, t_j) = 0 \tag{15b}$$

where the facet reflection is neglected.

2.3 Simulation procedure

So far, we have presented a detailed model of a single SOA. The simulation of the cascaded-SOA-switch can be completed by the following calculation procedure, where we use the case of 2 SOAs (turbo-switch) as the example:

1. Calculate the steady state of SOA1, and obtain the initial state of the carrier densities, ASE in each section, which will be used as initial conditions in following calculations. In the case of the steady state, the RHS of (1)-(3) should equal to zero, which implies that the carrier densities in each section will keep unchanging if the input does not change. A numerical algorithm from (Connelly, 2001) is adopted here to give a good convergence.
2. Calculate the response of SOA1 and get the output, by applying a proper input optical signal like a pump pulse or a pseudo-random binary sequence (PRBS) modulated pump pulse train, in addition to the probe CW beam. Firstly, according to initial conditions at time step t_1 obtained from step 1), carrier densities $N(z_i, t_2)$ can be calculated by using (11), so does $n_{CH}(z_i, t_2)$, $n_{SHB}(z_i, t_2)$, and the optical signal and ASE power in each section at the time step t_2 from (12-15). Thereby, all necessary quantities of SOA1 at time step t_2 are obtained, which can be treated as initial conditions to further calculations of the next time step. As the iteration completes, the N and P at each section and each time step can be obtained, which gives the output of the SOA1, $P(z_{Nz+1}, t_j)$.
3. Filter out the pump pulse or PRBS, and only allow the modulated CW signal to enter the SOA2.
4. Repeat step 1) to get the initial steady conditions of SOA2 firstly. It should be mentioned that, under this circumstance the amplified CW after SOA1 has to be used as the input to SOA2 to obtain the initial carrier densities and ASE levels in each section of SOA2.
5. Repeat step 2) using the modulated CW signal from the output of SOA1, as the input to SOA2. Calculate the output of SOA2, which is the final output of the turbo-switch.

3. Simulation results

The parameters used in our model are list in Table I. Two identical SOAs are applied in all the turbo-switch simulation, as implemented in the reported experiments. The SOAs are 0.7 mm long, which have a relatively high gain and short carrier lifetime, as well as an acceptable noise figure. A 200 mA bias current is consistently used unless specifically described. A 100% of the injected current utilization is supposed in the model. In the following simulations, the input CW and pump pulse are at wavelengths of 1560 and 1550 nm respectively, and the pulsewidth is 3 ps (FWHM) if not otherwise specified.

A steady state numerical algorithm presented in (Connelly, 2001) is used to obtain the SOA gain saturation characteristics, as illustrated in Fig. 3, where the wavelength of the CW input beam is 1560 nm. It is shown that, the small-signal gain of the amplifier is 25 dB, while the saturation output power is 12 dBm.

Symbol	Description	Value
L	Length of active region	0.7 mm
δ	Cross section area of active region	0.2 μm^2
Γ	Confinement factor	0.45
I	Injected / Bias current of the SOA	200 mA
A	Linear recombination coef.	21×10^{8} /s
B	Bimolecular recombination coef.	10×10^{-16} m^3/s
C	Auger recombination coef.	35×10^{-41} m^6/s
v_g	Group velocity in the active region	8.5×10^{7} m/s
N_0	Transparency carrier density	0.65×10^{24} /m^3
a_0	Differential gain	3.13×10^{-20} m^2
$\bar{a}$	Gain model coef.	1.2
b_0	Gain model coef.	3.17×10^{-32} m^4
b_1	Gain model coef.	0
λ_{p0}	Wavelength at peak	1575 nm
λ_{z0}	Wavelength at transparency	1625 nm
z_0	Gain model coef.	-2.5×10^{-33} m^4
τ_{CH}	Temperature relaxation time	700×10^{-15} s
τ_{SHB}	Carrier-carrier scattering time	70×10^{-15} s
ε_{CH}	Gain compression factor due to CH	1×10^{-23} m^3
ε_{SHB}	Gain compression factor due to SHB	0.5×10^{-23} m^3
α_{int}	Internal loss	5000
β_{ase}	Equivalent SE coupling factor	3.65×10^{-4}
λ_{ase}	Equivalent ASE wavelength	1550 nm
α_N	α-factor due to band-filling	8.0
α_T	α-factor due to CH	1.0

Table 1. SOA parameters used in the simulation

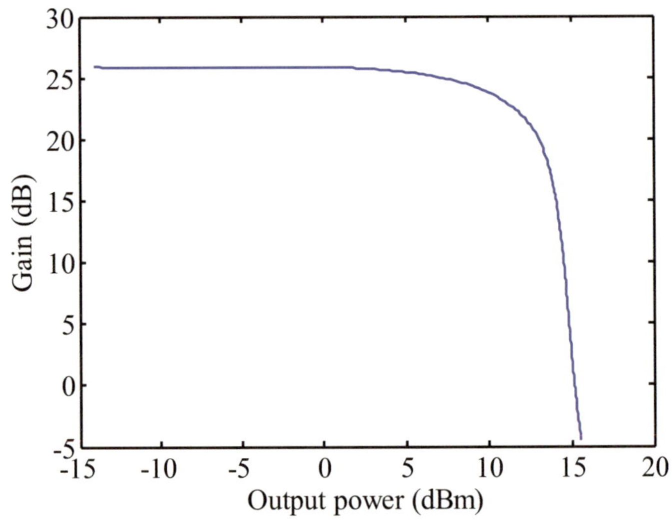

Fig. 3. Gain as a function of output power of a single SOA.

3.1 Gain and phase dynamics of turbo-switches

The gain dynamics of a single SOA and turbo-switch is plotted in Fig. 4. The input CW power is 0 dBm, while the pump pulse energy (single shot) is 100 fJ.

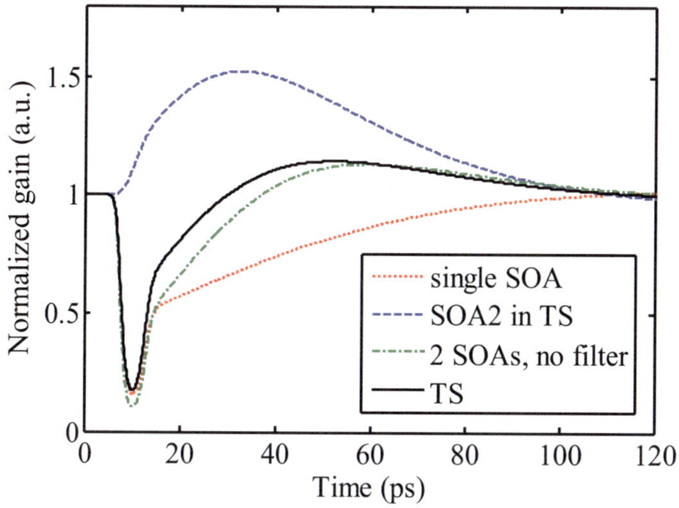

Fig. 4. Normalized gain of a single SOA, the SOA2 in TS, 2 cascaded SOAs with no filter between them, and the TS. TS: turbo-switch.

An obvious reduction of the gain recovery time is shown in the turbo-switch gain curve, comparing to the single SOA case, from about 100 ps to 20 ps, which is four times shorter than a single SOA. The simulation result is consistent with the corresponding experimental results presented in (Giller et al., 2006a). To get a better understand of the operating mechanism of turbo-switch, it is essential to know the gain response of SOA2, as plotted in Fig. 4. It is shown that, the gain curve of SOA2 has a completely different dynamics if compared with a single SOA. The gain of SOA2 increases firstly as the decrease of modulated CW input, and then starts to fall slowly back to the initial gain level. As a consequence, the slow recovery tail of the single SOA is somehow compensated, thus making the overall gain recovery of turbo-switch several times faster than that of a single SOA mechanism of turbo-switch, it is essential to know the gain response of SOA2, as plotted in Fig. 4. It is shown that, the gain curve of SOA2 has a completely different dynamics if compared with a single SOA. The gain of SOA2 increases firstly as the decrease of modulated CW input, and then starts to fall slowly back to the initial gain level. As a consequence, the slow recovery tail of the single SOA is somehow compensated, thus making the overall gain recovery of turbo-switch several times faster than that of a single SOA.

On the other hand, the phase dynamics curves are plotted in Fig. 5. It is shown that, turbo-switch also reduces the phase full recovery time from 100 ps to ~20 ps, about four times shorter than the case of a single SOA. It should be noted that the ultrafast effect of the SOA has much less impact on the phase change (Giller et al., 2006b), thus the phase recovery is mainly attributed to the inter-band processes, which makes it slightly different from the gain curve. To summarize, the turbo-switch scheme has shortened the overall gain/phase response time to a large scale compared with the case of a single SOA and has the capability of improving the overall operation speed of the switch to higher bit-rates.

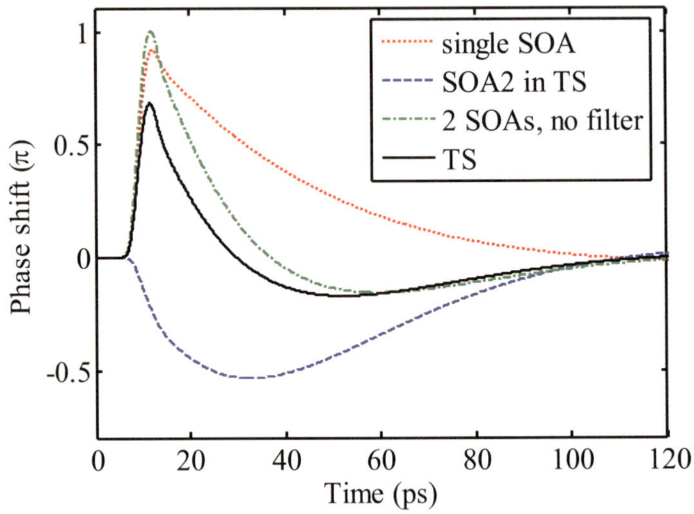

Fig. 5. Normalized phase dynamics of a single SOA, the SOA2 in TS, 2 cascaded SOAs with no filter between them, and the TS. TS: turbo-switch.

Moreover, our simulation shows that filtering the pump pulses before the SOA2 in turbo-switch scheme does further reduce the gain/phase recovery time, when compared to the case of no filter between two SOAs (dash-dotted gain/phase curves in Fig. 4), as presented in (Marcenas et al., 1995). In the latter case, the pump was entered into two cascaded SOAs along with the probe CW beam, which can also be regarded as a single long SOA with a double length of the active waveguide.

3.2 Pattern-effect mitigation of turbo-switches

As a result of the shorter gain/phase recovery time in turbo-switch, the pattern effect associated with the slow recovery is supposed to be mitigated. It should be noted that, along with the faster gain response of the turbo-switch, an overshoot in the gain/phase curve can also be clearly observed. However, the overshoot level can be controlled by adjusting the average input optical power to the SOA2. To verify the mitigation of the pattern effect, a variable optical attenuator (VOA) is experimentally applied before SOA2 in the turbo-switch configuration to optimize the output pattern of the turbo-switch (Giller et al., 2006c).

The simulation results of the turbo-switch gain dynamics under a single shot of 3 ps pump pulse are presented in Fig. 6, as a function of power levels before SOA2, which are in good agreement with the experimental results (see Fig. 6(b)) presented in (Giller et al., 2006c). In the simulation, the input CW power to SOA1 is 0 dBm and pump pulse energy is 100 fJ. It is shown that, when reducing the input power to SOA2, the recovery time becomes longer, and the level of the overshoot is lower. When the input power becomes low enough to make SOA2 unsaturated, the overall turbo-switch gain response will exhibit similar to that of a single SOA. So there is an optimum input power level for SOA2 in order to achieve the optimum effective recovery time at a specific data bit-rate.

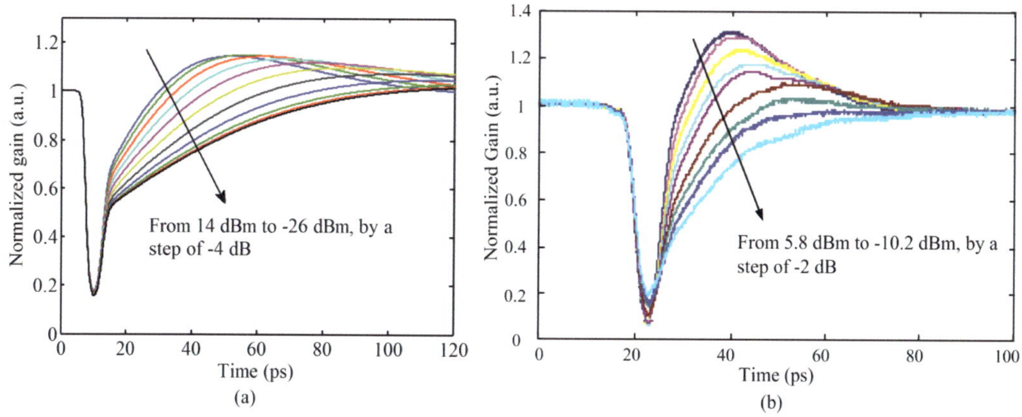

Fig. 6. Normalized gain dynamics of turbo-switch with a VOA before SOA2. (a).Simulation, where the optical power input to SOA2 varies from 14 to -26 dBm, by -4 dB each step. (b). Experimental result, where the optical power input to SOA2 varies from 5.8 to -10.2 dBm by a step of -2 dB (Giller et al., 2006c).

To show the pattern effect of turbo-switch, the output patterns of a CW probe beam modulated by a 40 Gb/s PRBS pump pulse train are presented in Fig. 7(a)-(c), where three different input power levels are chosen before SOA2: -26, -11, and 14 dBm. The input CW power before SOA1 is 0 dBm and the pump pulse energy is 2 fJ.

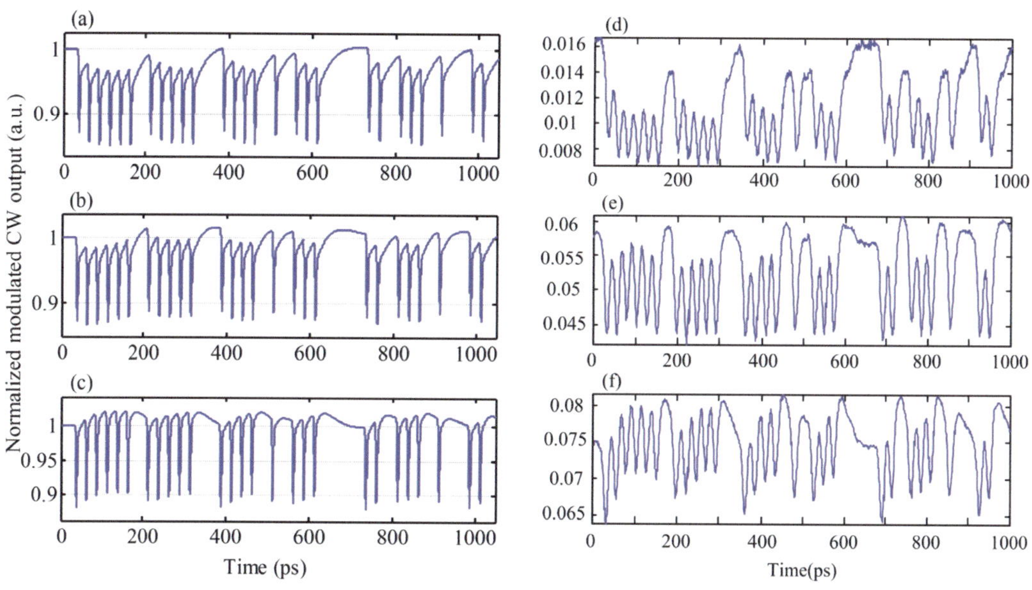

Fig. 7. Modulated CW output patterns from the TS, with a PRBS pump pulse train at 40 Gb/s. (a)-(c) Simulation, where, the optical power input to SOA2 in simulation are -26, -11 and 14 dBm respectively from top to bottom. (d)-(f). Experimental results (Giller et al., 2006c).

It is shown in Fig. 7 that, the simulation results are in good agreement with the experimental measurements (see Fig. 7(d)-(f)) presented in (Giller et al., 2006c) for all three cases.

Consequently, there is an optimum optical power level that could mitigate the pattern effect at a specific bit-rate. For instance, in this case, the optimum power to SOA2 should be -11 dBm, as shown in Fig. 7(b), where the pattern effect is mitigated.
On the contrary, the simulated modulation curves in Fig. 7(a) and 7(c) both experience a worse trend (constantly lower or higher) under a sequence of consecutive marks (ones) or spaces (zeros). This implies that, the unsaturated (-26 dBm), saturated (-11 dBm) and over-saturated (14 dBm) input power level to SOA2 has an important impact on the overall performance of turbo-switch. For instance, in the case of the modulated CW power of -26 dBm, the SOA2 cannot be saturated, and the overall recovery time is not shortened, turbo-switch behaviors similarly to a single SOA, as shown in Fig. 7(a).

4. Applications of turbo-switch

The turbo-switches are supposed to be applied in the all-optical signal processing in order to enhance the operation speed. The turbo-switches have been employed as the all-optical AND gate (wavelength conversion) and XOR gate, whose operation speeds have been increased up to 160 Gb/s and 85 Gb/s respectively. In this section, we will demonstrate the details of the high-speed operation of the turbo-switches, from both theoretical and experimental aspects.

4.1 High-speed AND gate beyond 160 Gb/s

The simulation results of the AND gate at 160 Gb/s will be given in Section 4.1.1, while the corresponding experimental results (i.e., wavelength conversion) at 170 Gb/s will be presented in Section 4.1.2.

4.1.1 Simulation of AND gate based on turbo-switch

A simulation was carried out to evaluate the 160 Gb/s all-optical wavelength conversion using a turbo-switch and a delayed interferometer (DI). The setup is shown in Fig. 8(a), where a polarization maintaining fiber (PMF) and a polarizer are used to form a DI (Reid et

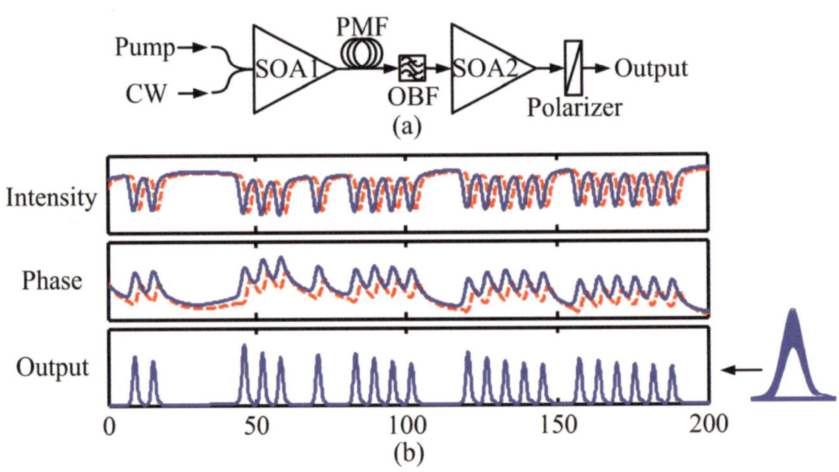

Fig. 8. All optical wavelength conversion using a turbo-switch. (a) Setup; (b) 160 Gb/s simulation results. The blue/ red curves are the TE/TM polarized components. Note that polarization controllers are not plotted for simplicity.

al., 2008). The DI utilizes the differential cross phase modulation (XPM) effect of the SOA to achieve the polarity-maintaining wavelength conversion. The PMF is used to introduce a delay between the TE / TM components of the probe, thus introducing a differential phase shift between the two orthogonal components. On the other hand, the polarizer acts as an interfering device to extract phase difference between the two components, which is the wavelength-converted output.

Fig. 8(b) presents the 160 Gb/s wavelength-converted output trace and the corresponding eye diagram. The average powers of the CW and the pump are 10 and 3 dBm respectively, while the wavelengths of the CW and pump are 1560 and 1550 nm respectively. The input PRBS data has a length of 2^7-1. The PMF gives a differential delay of 2 ps. It is shown that the turbo-switch configuration expedites the recovery of the intensity and phase, which helps to mitigate the patterning of the output. The clearly opening eye diagram of the output shows the feasibility of the wavelength conversion at 160 Gb/s. More specifically, the well-known Q factor, defined for instance in (Agrawal, 2002), for the output signal is 6.8, which corresponds to the bit error rate (BER) of 6.9×10^{-12}.

4.1.2 Experiment of AND gate based on turbo-switch

The wavelength conversion incorporating a turbo-switch was experimentally verified using the setup shown in Fig. 9(a), at ~85 and 170 Gb/s. The wavelength converter had the configuration of the DI. In a DI, CW light is amplitude and phase modulated in SOA1 by the action of the data pulse stream, and is then split into 'fast' and 'slow' components that travel along the two axes of a length of PM fiber. The two components experience a differential delay, Δt (3ps, in our experiment). The phase difference between them results in a polarization rotation when they interfere at the polarizer, and hence switching of the CW beam occurs, with a non-inverted output. The wavelength-converted output was de-multiplexed down to 42.6 Gb/s using MZ modulators.

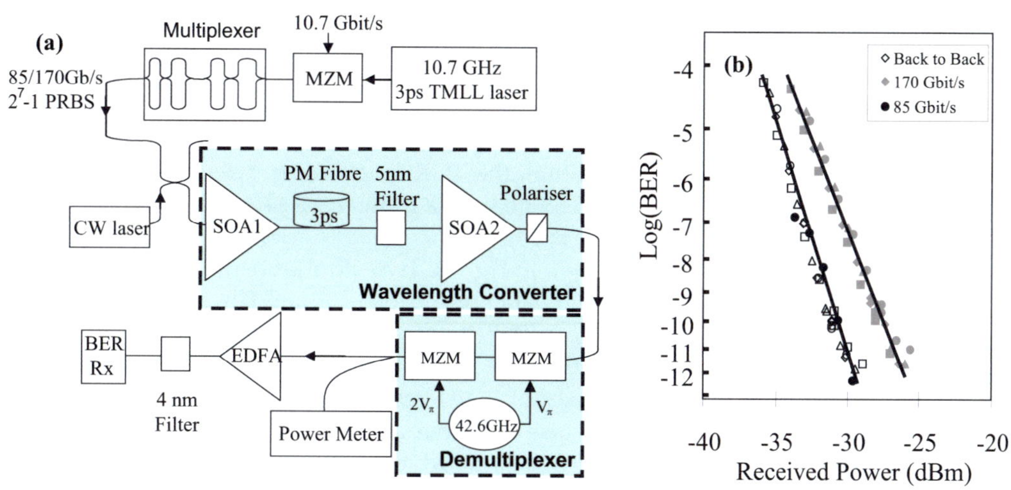

Fig. 9. 170 Gb/s wavelength conversion using DI configuration incorporating a turbo-switch. (a). The setup; (b). BER curves demultiplexed to 42.6 Gb/s for back-to-back and wavelength-converted signals at 85 and 170 Gb/s (Manning et al., 2006).

Fig. 9(b) shows the BER curves for the 42.6 Gb/s channels for wavelength conversion at 85 and 170 Gb/s. We observed no power penalty at 85 Gb/s, and a 3 dB penalty at 170 Gb/s, which we believe was due to the pulse width of the converted channels of 3 ps being slightly too long for 170 Gb/s data, and implied that our differential delay Δt is non-optimal. The measured OSNR was 40 dB, referred to a 0.1 nm noise bandwidth.

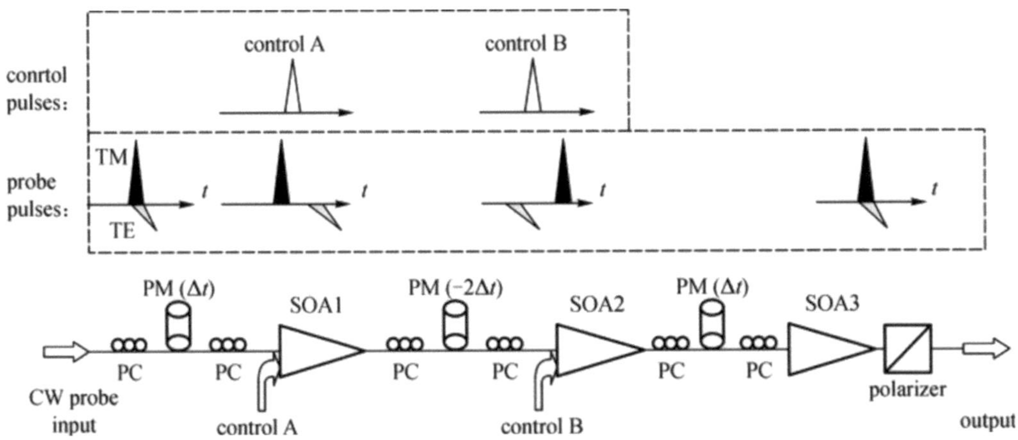

Fig. 10. Principle of dual-ÚNI XOR logic gate (PM: polarization maintaining fiber; PC: polarization controller)

4.2 High-speed XOR gate based on turbo-switch

Section 4.2.1 gives operation principle of the 85 Gb/s XOR gate, while the experimental results including eye diagram and the spectrum of the output are present in Section 4.2.2.

4.2.1 Principle of all-optical XOR gate

All-optical XOR logic is regarded as one of the fundamental logic gates in signal processing, which plays an important role in applications such as bit pattern recognition (Webb et al., 2009), pseudorandom bit sequence (PRBS) generation, parity checking and optical computing. A high-speed all-optical XOR gate has potential applications for on-the-fly digital serial processing of optical signals, for example, in packet header recognition, error detection and coding/decoding. In addition, the XOR function has been used recently as a wavelength converter and regenerator for signals in a differential phase shift keying (DPSK) format (Sartorius et al., 2006; Kang et al., 2005).

The scheme of the XOR logic gate is shown in Fig. 10. Two ultrafast nonlinear interferometer (UNI) elements are cascaded to allow two data pulse streams A and B to be input into SOAs 1 and 2 respectively as control pulses. The input probe pulses are launched into a polarization-maintaining (PM) fiber with equal intensities on the fast and slow fiber axes, which are coupled to the TM and TE axes of SOA1 respectively. As a result, the TE pulse lags the TM pulse by Δt. The control pulse A is introduced between the two probe pulse components before they are input into SOA1, in which it induces a π-radian phase shift experienced by the TE pulse alone. The probe pulses are then injected into another PM fiber with a differential delay of $-2\Delta t$. The fast and slow axes of this PM fiber are orthogonal to those of the first section, resulting in a reversal of the delay between TE and TM pulses so that the TE pulse is now Δt ahead of the TM pulse. The control pulse B is then introduced

between the TE and TM probe pulses before entering SOA2 where now the induced π-radian phase shift affects only the TM pulse. The third PM fiber, with differential delay, Δt, resynchronizes the TE and TM probe pulses in time. A π-radian phase shift between TE and TM pulses gives a polarization rotation of $\pi/2$ when they recombine at the polarizer, which is crossed with respect to the un-rotated probe.
When both of the control pulses A and B are present, the nonlinear phase difference between TE and TM will be zero (first π, then - π), the same result as the case when both A and B are absent. In the cases of either A or B alone being present, the phase shift will be ±π. The system is biased OFF (no output) in the absence of the control pulses. A pulse is generated after the polarizer only when one of A and B is present. Thus the operation of the device satisfies the XOR logic truth table as shown in Table 2. As with the conventional UNI gate, the probe pulses may be replaced by a continuous wave (CW) beam, in which case the output takes the form of pulses of width Δt.

Data A	Data B	XOR
0	0	0
0	1	1
1	0	1
1	1	0

Table 2. Truth table of XOR gate

The transmission of control pulse A is blocked by a filter (not shown) placed before SOA2. SOA1 and SOA2 are therefore configured as a turbo-switch (Manning et al., 2006) and the effective switching speed by control pulse A is enhanced. The addition of a third SOA (also after a filter) to the original dual ultrafast nonlinear interferometer XOR gate (DUX) forms a second turbo-switch that similarly enhances the speed of switching by control pulse B.

4.2.2 Experiment of XOR gate based on turbo-switch

An experiment was carried out to evaluate the performance the proposed XOR scheme at 85 Gb/s, whose experimental setup is shown in Fig. 11. A CW laser with a wavelength of 1552 nm was employed as the probe beam instead of a pulse train, so the first PM fiber in Fig. 10 was not required. The 3 ps, 1557 nm control pulses A and B were obtained from a 10.645 GHz mode-locked laser The control pulse stream was optically modulated with a 2^7-1 pseudo-random bit sequence (PRBS) and the pulses were passively multiplexed to 85 Gb/s before being injected into the SOAs 1 and 2. An optical delay-line was used to present different parts of the sequence to each SOA. Two variable optical attenuators (VOAs) were employed to adjust the control pulse energies. Another VOA was used to optimize the input power of the probe beam injected to SOA2.
All the three (Kamelian) SOAs were biased at 400mA, where their unsaturated gain was greater than 30 dB. The differential delays of PM fibers were 11.5 ps ($2\Delta t$) and 5.75 ps (Δt) respectively, where Δt is one half of the bit period at 85 Gb/s. 5 nm band-pass filters blocked the control pulses and allowed the propagation of the probe beam. The polarization controllers (PC) in front of each PM fiber were adjusted to launch approximately equal amplitudes into the TE and TM modes. The two polarization states were also aligned with the TE and TM modes of the active layer at the input to SOAs 1 and 2 with further PCs. This was to prevent the control pulses causing polarization rotation.

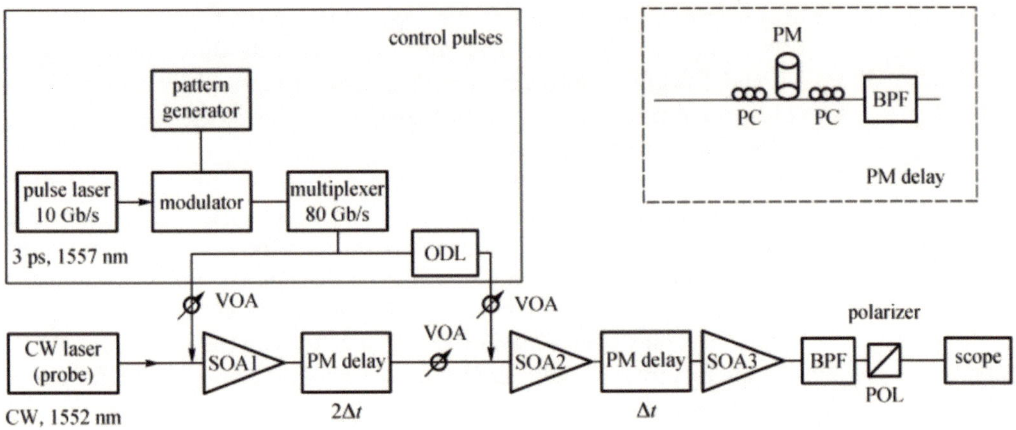

Fig. 11. Experimental setup of 85-Gb/s XOR logic gate, where PM delay indicates a length of PM fiber with PCs and a filter (inset).

The output of the XOR logic gate was monitored by a 70 GHz oscilloscope. XOR operation was realized at 10, 21, 42 and 85 Gb/s by adjusting the control pulse multiplexer. The amplitude variations in the 85 Gb/s output eye diagram (Fig. 12) were primarily due to imperfections in the multiplexer. The output spectrum at the same rate is shown in Fig. 13, where the sidebands are visible, but suppressed compared to a normal return-to-zero AND gate spectrum (The inset of Fig. 13). This is because the output pulses resulting from an $A \bullet \overline{B}$ input are in anti-phase to those corresponding to $B \bullet \overline{A}$. The average powers of the probe beam were 4 dBm before SOA1 and SOA2 and 10 dBm before SOA3. The average powers of control pulses A and B were 4 dBm and 3.5 dBm respectively, implying control pulse energies of 54 fJ and 62 fJ.

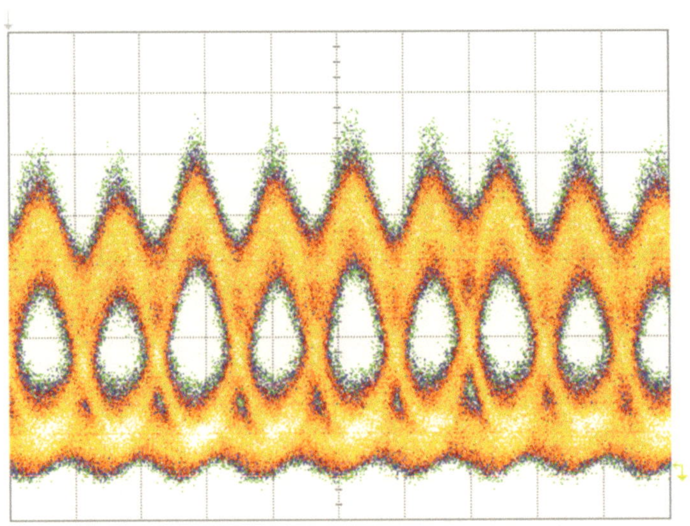

Fig. 12. 85 Gb/s XOR output eye diagram (5 ps/division).

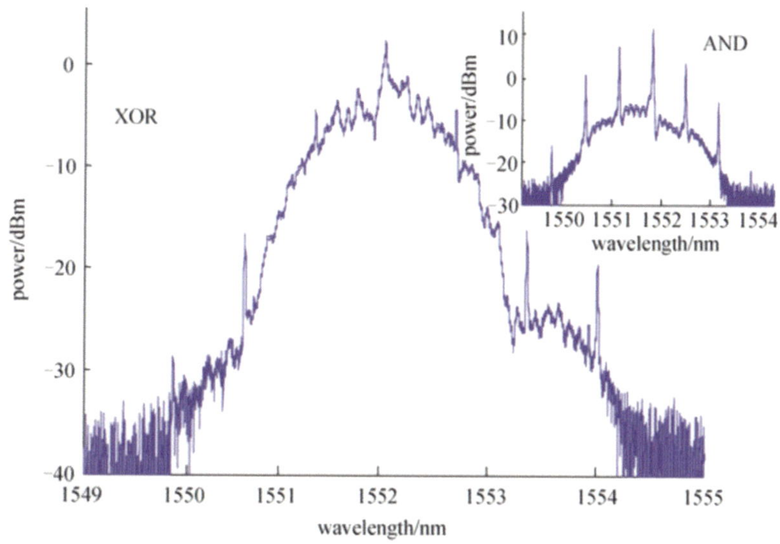

Fig. 13. Spectrum of 85 Gb/s XOR output (resolution is 0.01 nm; the inset is the corresponding spectrum of an 85 Gb/s AND gate)

5. Further improvement of the switch speed

In the turbo-switch structure, an extra SOA2 is cascaded following the SOA1, acting as a high-pass filter to filter out the slow response associated with the SOA carrier lifetime. In such a way, the overall operation speed of the turbo-switch device has been demonstrated several times faster than that of a single SOA. A straight-forward idea is that, if more SOAs are cascaded after the turbo-switch, is there any further improvement to the operation speed?

5.1 Switch of multiple cascaded SOAs

For the multiple cascaded SOAs, the simulations are carried out. The gain recovery time, overshoot level (normalized by the initial power level), and noise figure as a function of SOA stage are plotted in Fig. 14, where the powers of input CW probe and pump pulse to SOA1 are the same as Fig. 4. The noise figures of the cascaded switches are obtained using the equations presented in (Baney et al., 2000).

The results are actually encouraging, since the recovery time is reduced to ~10 ps when three SOAs are cascaded, which implies that more SOAs after turbo-switch, faster recovery could be expected. However, the degree of the overshoot and noise figure also rise almost linearly as the numbers of SOA increases, whereas the recovery time is not reduced significantly any more after the stage number exceeding 5. Moreover, the ASE noise and the complexity of the device are also expected to increase when more SOAs are cascaded. Therefore a trade-off has to be considered accordingly when choosing an optimum structure of turbo-switch for a specific application. Nevertheless, our simulation suggests that, the optimum number of SOA should be in the range of 2 to 5.

It should be mentioned that, SOAs can be cascaded directly without any filter between them as presented in (Marcenac & Mecozzi, 1997), where it requires ten SOAs to achieve a speed of 110 Gb/s, whereas our simulation shows that the same operation speed can be achieved with a three-SOA-switch if a filter is implemented, as indicated in Fig.14.

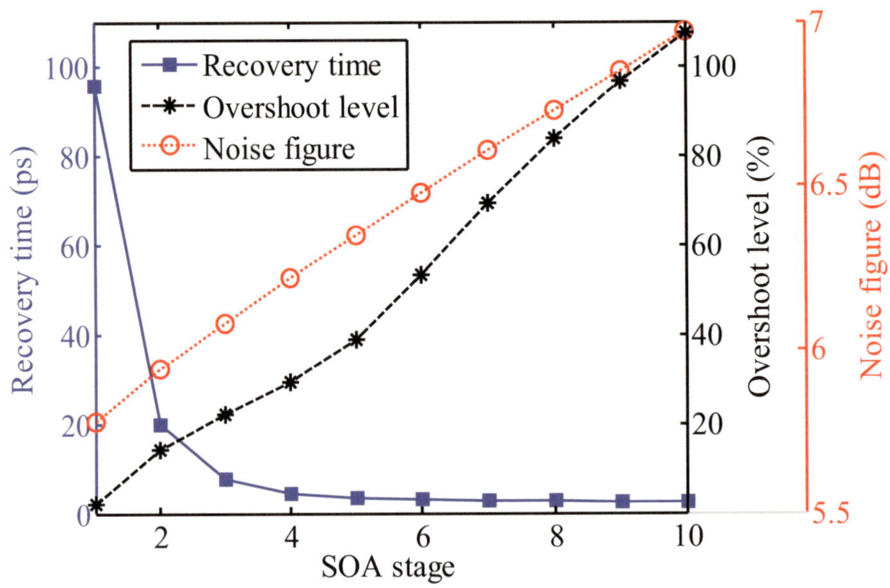

Fig. 14. Gain recovery time, overshoot level, and noise figure as a function of the number of the SOA stages.

5.2 Impact of SOA injected currents

For all the figures presented before this subsection, the SOA currents are fixed at 200 mA. Since the pattern effect can be mitigated using a VOA before the SOA2, as shown in Fig. 6. One could consider that whether the variation of the SOA current level can introduce a similar effect as the VOA? The simulation is carried out by varying one of the injected current of the SOA1 and SOA2 in turbo-switch scheme.

As shown in Fig. 15(a) and 15(b), the current variation curves reveal a similar effect as that of Fig. 6, where the nonlinearity of the SOA2 is gradually diminishing as the current keep decreasing, which consequently results in the overall response of turbo-switch similar to that of a single SOA. However, an interesting phenomenon is that, the gain dynamic is quite different when the current of SOA1 is reduced, as shown in Fig. 15(a). In the latter case, even though an overshoot is obvious when the current level is high, the full recovery time is generally longer than the case of Fig. 15(b). Apart from that, the gain compression induced by the pump pulse is smaller as well, which will potentially affect the extinction ratio (ER) of the output signal. To summarize, it is better to set the current level of SOA1 high, while the current level of SOA2 can be employed to optimize the overall gain/phase recovery time of the turbo-switch.

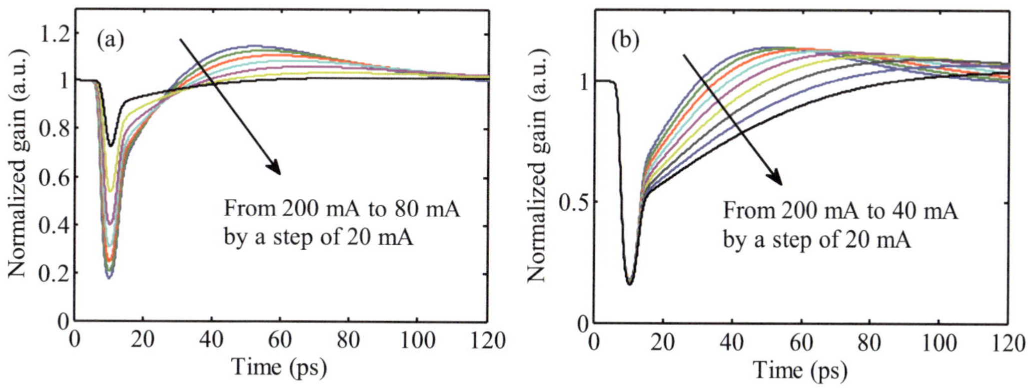

Fig. 15. Normalized gain dynamics of turbo-switch, by varying the injected current level of (a) SOA1, (b) SOA2.

6. Conclusions

A detailed theoretical model to simulate the gain/phase dynamics of the input optical signals propagating through turbo-switch has been presented. The simulation results have been shown in excellent agreement with experimental measurements incorporating the turbo-switch, in terms of CW modulation, pattern effect mitigation at 40 Gb/s, and eye diagram of 160 Gb/s wavelength conversion based on turbo-switch. The introduction of turbo-switch has the capability of increasing the overall switch speed by a factor of four if compared with the case of a single SOA switch, which is also confirmed by the previous experimental demonstration.

Moreover, the theoretical analysis based on the SOA model suggests that, higher bit-rate operation can be expected, if using an extended turbo-switch structure for instance with three or more cascaded SOAs. In addition, optimized configurations of turbo-switch with the differential XPM scheme as well as the bias current of SOAs can also be implemented to achieve a potential higher switch speed.

7. Acknowledgments

The authors would like to thank our colleagues Dr. R. J. Manning, Dr. R. P. Webb, and Dr. R. Giller from Tyndall National Institute, Ireland, for the experiment facilities and results used in the paper for comparisons.

8. References

Agrawal G. P. and Olsson N. A. (1989). Self-phase modulation and spectral broadening of optical pulses in semiconductor laser amplifiers. *IEEE J. Quantum Electron.*, vol. 25, no.11, pp. 2297-2306, ISSN 0018-9197.

Agrawal G. P. (2002). *Fiber-optic communication systems* (3rd edition), Wiley-Interscience, Wiley series in microwave and optical engineering, ISBN 978-0471215714,USA.

Baney D. M., Gallion P. and Tucker R. S. (2000). Theory and measurement techniques for the noise figure of optical amplifiers. *Opt. Fiber Technol.*,vol. 6, no. 2, pp. 122–154, ISSN 1068-5200.

Bischoff S., Nielsen M. L. and Mørk J. (2004). Improving the all-optical response of SOAs using a modulated holding signal. *J. Lightwave Techonol.*, vol. 22, no.5, pp. 1303-1308, ISSN 0733-8724.

Connelly M. J. (2001). Wideband semiconductor optical amplifier steady-state numerical model. *IEEE J. Quantum Electron.*, vol. 37, no. 3, pp. 439-447, ISSN 0018-9197.

Dutta N. K. and Wang Q. (2006). *Semiconductor optical amplifier*, World Scientific, ISBN 981-256-397-0, Singapore.

Giller R., Manning R. J. and Cotter D. (2006a). Recovery dynamics of the "turbo-switch". *Optical Amplifiers and Their Applications (OAA)*, Whistler, Canada, Jun. 2006. Paper OTuC2.

Giller R., Manning R.J. and Cotter D. (2006b). Gain and phase recovery of optical excited semiconductor optical amplifier. *IEEE Photon. Technol. Lett.*, vol. 18, no. 9, pp. 1061-1063, ISSN 1041-1135.

Giller R., Yang X., Manning R. J., Webb R. P. and Cotter D. (2006c). Pattern effect mitigation in the turbo-switch. *International Conference on Photonics in Switching*, Heraklion, Crete, Oct. 2006.

Gutiérrez-Castrejón R. (2009). Turbo-switched Mach-Zehnder interferometer performance as all-optical signal processing element at 160 Gb/s. *Opt. Commun.*, vol. 282, no. 22, pp. 4345-4352, ISSN 0030-4018.

Kang I., Dorrer C., Zhang L., Rasras M., Buhl L., Bhardwaj A., Cabot S., Dinu M., Liu X., Cappuzzo M., Gomez L., Wong-Foy A., Chen Y. F., Patel S., Neilson D. T., Jaques J. and Giles C. R. (2005). Regenerative all optical wavelength conversion of 40 Gb/s DPSK signals using a semiconductor optical amplifier Mach-Zehnder interferometer. *ECOC*, Glasgow, Scottland, Sep. 2005. Paper Th4.3.3.

Leuthold J., Mayer M., Eckner J., Guekos G., Melchior H. and Zellweger Ch. (2000). Material gain of bulk 1.55 mm InGaAsP/InP semiconductor optical amplifiers approximated by a polynomial model. *J. App. Phys.*, vol. 87, no. 1, pp. 618-620, ISSN 0021-8979 .

Leuthold J. (2002). Signal regeneration and all-optical wavelength conversion. *Annual Laser and Electro Optics Society (LEOS) Meeting 2002*, Glasgow, Scottland, Nov. 2002. Paper MM1.

Liu Y., Tangdiongga E., Li Z., Zhang S., Waardt H. de, Khoe G. D., and Dorren H. J. S. (2006). Error-free all-optical wavelength conversion at 160 Gb/s using a semiconductor optical amplifier and an optical bandpass filter. *J. Lightwave Technol.*, vol. 24, no.1, pp. 230-236, ISSN 0733-8724.

Liu Y., Tangdiongga E., Li Z., Waardt H. de, Koonen A. M. J., Khoe G. D., Shu X., Bennion I. and Dorren H. J. S. (2007). Error-free 320-Gb/s all-optical wavelength conversion using a single semiconductor optical amplifier. *J. Lightwave Techonol.*, vol. 25, no.1, pp. 103-108, ISSN 0733-8724.

Manning R. J., Yang X., Webb R. P., Giller R., Gunning F. C. G. and Ellis A. D. (2006). The turbo-switch—a novel technique to increase the high-speed response of SOAs for wavelength conversion. *OFC/NFOEC,* Anaheim, CA., Mar. 2006. Paper OSW8.

Manning R. J., Giller R., Yang X., Webb R. P. and Cotter D. (2007). SOAs for All-optical switching-techniques for increasing the speed. *International Conf. on Transparent Optical Networks,* Rome, Italy, Jul. 2007. pp. 239-242.

Marcenas D. D., Kelly A. E., Nesset D. and Davies D. A. O. (1995). Bandwidth enhancement of wavelength conversion via cross-gain modulation by semiconductor optical amplifier cascade. *Electron. Lett.,* vol. 31, no. 17, pp. 1442-1443, ISSN 0013-5194.

Marcenac D. and Mecozzi A. (1997). Switches and frequency converters based on cross-gain modulation in semiconductor optical amplifiers. *IEEE Photon. Technol. Lett.,* vol. 9, no. 6, pp. 749-751, ISSN 1041-1135.

Mecozzi A. and Mørk J. (1997). Saturation effects in non-degenerate Four-Wave mixing between short optical pulses in semiconductor laser amplifiers. *IEEE J. Sel. Topics Quantum Electron.,* vol. 3, no.5, pp. 1190-1207, ISSN 1077-260X.

Poustie A. (2007). SOA-based all-optical processing. *OFC/NFOEC.* Anaheim, CA., Mar. 2007. Tutorial-OWF1

Reid D. A., Clarke A. M., Yang X., Maher R., Webb R. P., Manning R. J. and Barry L.P. (2008). Characterization of a turbo-switch SOA wavelength converter using spectrographic pulse measurement. *IEEE J. Sel. Topics Quantum Electron.,* vol. 14, no. 3, pp. 841-848, ISSN 1077-260X.

Sartorius B., Bornholdt C., Slovak J., Schlak M., Schmidt C., Marculescu A., Vorreau P., Tsadka S., Freude W. and Leuthold J. (2006). All optical DPSK wavelength converter based on MZI with integrated SOAs and phase shifters. *OFC/NFOEC,* Anaheim, CA., Mar. 2006. Paper OWS6.

Stubkjaer K.E. (2000). Semiconductor optical amplifier-based all-optical gates for high-speed optical processing. *IEEE Journal of Selected Topics in Quantum Electronics,* vol.6, no.6, pp1428~1435, ISSN 1077-260X.

Talli G. and Adams M. J. (2003). Gain dynamics of semiconductor optical amplifiers and three-wavelength devices. *IEEE J. Quantum Electron.,* vol. 39, no.10, pp. 1305-1313, ISSN 0018-9197.

Webb R. P., Yang X., Manning R. J., Maxwell G. D., Poustie A. J., Lardenois S., Cotter D. (2009). All-optical binary pattern recognition at 42 Gb/s. *Journal of Lightw. Technol.,* vol. 27, no. 13, pp. 2240–2245, ISSN 0733-8724.

Yang X., Lenstra D., Khoe G.D. and Dorren H.J.S. (2003). Nonlinear polarization rotation induced by ultra-short optical pulses in a semiconductor optical amplifier, *Optics Comm.,* vol.223, no.1-3, pp.169-179, ISSN 0030-4018.

Yang X., Manning R. J. and Webb R. P. (2006). All-optical 85Gb/s XOR using dual ultrafast nonlinear interferometers and turbo-switch configuration. *ECOC,* Cannes, France, Sep. 2006. Paper Th1.4.2.

Yang X., Weng Q. and Hu W. (2010). High-speed, all-optical XOR gates using semiconductor optical amplifiers in ultrafast nonlinear interferometers. *Front. Optoelectron. China*, vol. 3, no. 3, pp.245–252, ISSN 1674-4128.

3

A Novel Method of Developing Frequency Encoded Different Optical Logic Processors Using Semiconductor Optical Amplifier

Sisir Kumar Garai
Department of Physics, M.U.C. Women's College, Burdwan, West Bengal, India

1. Introduction

To implement different digital processors in optical domain, encoding and decoding of optical data are the prime issues. Till now several encoding/decoding techniques have been reported for representing the optical information. In this connection spatial encoding [Toyohiko Y., 1986], intensity encoding [Mukhopadhyay S.,et-al., 2004], polarization encoding [Awwal A.A.S., et-al., 1990; Zaghloul Y.A., et-al., 2006, 2011], phase encoding [Chakraborty B., et-al., 2009] etc. may be mentioned. But these coding processes have some inherent problems. In spatial encoding, two specific pixcells, each having two different types of opaque and transparent sub-cells distribution are encoded either as '1' and '0'states respectively in 2-D plane. Here input signal bits are generated by electro-optic/electronic switching (with suitable nonlinear materials) which limits the speed of processing. Again in pixels based operation, interference and diffraction effect may change the expected result of the image pattern at the output end which leads to bit error problem. Moreover, as output result is obtained using decoding mask, and the encoding and decoding technologies not being the same, therefore it is not possible to design sequential or combinational logic circuit using spatial encoding technique. In intensity encoding, presence of optical signal or the intensity of a signal greater than that of a specific reference intensity have been encoded as '1' state and absence of signal or the intensity of a signal lower than that of a specific reference intensity have been encoded as '0' state. But for long distance communication, intensity of optical signal may fall and dropdown below the reference level and for which the '1' state may be treated as '0'state of the signal which can also lead to the bit error problem. In most of the cases the all-optical logic gates are implemented by non-linear materials extending its 2nd order of nonlinearity. This material sends the light passing through it in different channels if the intensity of light varies. So the change of a prefixed value of intensity creates some major problems in channel selection and therefore this intensity based encoding principle is problematic. In intensity based refractive index variation technology, small fluctuation of intensity of the input beams may collapse the total set up. In polarization encoding, one specific state of polarization of the optical beam is encoded as '0' state and another specific orthogonal state of polarization is treated as '1' state. Again, the state of polarization may change for several causes which can also lead to the bit error in information processing. In phase encoding, one specific phase of the optical beams is encoded as '0' state and another specific phase is treated as 1 state. But it is very difficult to maintain the constant phase relationship throughout the optical signal processing, specially, beyond the coherent length. Similarly the other coding norms may extend some other limitations in wide range data processing.

In contrast to the above mentioned encoding, the author has established the frequency encoded technique to represent the Boolean logic states. It is known that if '1' and '0' logic states are encoded by two different frequencies in optical domain, then one may ensure about the state of a signal during data transmission. If '0' state is encoded by the frequency 'v_1'and the '1'state by the other frequency 'v_2' then 'v_1' and 'v_2' will normally remain unaltered throughout the transmission of data. The frequency encoded technique offers so many advantages [Garai S.K., Mukhopadhyay S.(2009),2010; Garai S.K.(2010); Garai S.K.(2011),2011a,2011b]. The prime beauty of the frequency encoding is that, frequency is the fundamental property of the wave and it can preserve its identity irrespective of the absorption, reflection, transmission during its propagation throughout the communicating media. This is the most potential advantage of the frequency encoding technique over other conventional encoding techniques. In addition, frequency encoding in optical domain uses the spectrum of a broadband optical source and can accommodate a large pool of subscribers. Moreover different signals are characterized by different specific frequencies in optical domain and if one signal of specific frequency can be encoded to represent a specific state of information, then using different signals of different frequency, other different states can be encoded. Thus a larger number of states of information can be accommodated which can propagate through the same channel i.e., through the same optical fiber without interference or cross-talk. Again using frequency encoding it is easier to represent multi-bit states of information which are very useful for conducting multivalued logic operations using wavelength conversion properties of different high speed nonlinear optical switches such as semiconductor optical amplifier, periodically poled lithium niobate ($LiNbO_3$) waveguide, Chalkogenide glass etc. Since the information is frequency encoded, therefore the coded signal is very useful for optical wavelength division multiplexing (WDM), frequency division multiplexing (FDM) and combination of WDM and time division multiplexing (TDM) in interconnection of telecommunication networks.

Basic building blocks required to implement the frequency encoded optical logic processors are the Frequency Router Unit (F.R.U) and Frequency Converter Unit (F.C.U) and SOA is found to be the very promising in this aspect. A rapid growth in the optical fibre communication was noticed over the last thirty years exploiting the enormous bandwidth property and many other characters of optical fibre. The massive advancement of optical technology has been made possible because of several reasons. In this regard it can be specially mentioned that Semiconductor Optical Amplifier (SOA) is a promising optical device that help a lot for the acceleration of advancing the network systems in communication. SOAs are highly nonlinear in an optical gain range. This is due to the consequence of a large number of free carriers confined in a small active region and it affects the gain as well as refractive index within the active region. The SOA nonlinear properties such as cross gain modulation (XGM), cross phase modulation (XPM), four-wave mixing (FWM) have been studied several times and are applied to implement wavelength conversion, optical division multiplexing-demultiplexing, clock recovery, and optical logic gates[Connelly M.J.(2002); Dutta N.K. et.al.,2006, Asghari M.,et-al.,1997; Soto.H.,et.al.,1999; Guo L.Q., Connelly M.J.(2007)]. The wavelength conversion by XGM is accompanied by large chirp and low extinction ratio with restricted speed (limited by the carrier recovery time) up to 40Gbit/s and even up to 100Gbit/ with some degradation. On the other hand the XPM schemes enable wavelength conversion with lower signal powers, reduced chirp, enhanced extinction ratios and ultra fast speed of switching. The wavelength conversion by FWM is very promising one due to its independent modulation format as well as dispersion compensation property with ultra speed However, it has polarization sensitive low wavelength conversion efficiency. Wavelength conversion based on cross polarization modulation (XPolM) is another promising approach. In XPolM process, nonlinear polarization rotation (NPR), an optically induced birefringence and dichroism property of an SOA have been exploited for wavelength conversion and it has drawn the attention of scientists and technologists [Guo L.Q., Connelly M.J.(2005,2006); Lacey J.P.R., et.al.,1998] Very recently all-optical wavelength (both up and down) conversion has realized exploiting non-linear properties of SOA [Guo L.Q., Connelly M.J.(2008)]. In this chapter the author has presented a novel method of developing all optical logic processor exploiting frequency conversion and switching character of SOA. The author has organized the chapter in brief based on his established works and actually it is a review one with some modifications.

The Chapter covers (a) a method of generating all-optical decimal data to frequency encoded binary data (b) a method of developing all optical frequency encoded binary logic gates such as AND, OR, NAND, NOR, EX-OR and finally (c) an all optical memory unit, and all of these are the integral part of the all-optical logic processors and these are developed exploiting different attractive features of SOA. The author has exploited here the principle of nonlinear rotation of the state of polarization (SOP) of the probe beam in semiconductor optical amplifier for the frequency conversion as well as for the switching purpose and this type of switching is so called polarization switching (PSW). The chapter is organized as follow: In section-2, the author has presented the basic principle of frequency conversion using nonlinear polarization rotation (NPR) of the probe beam in SOA, principle of channel (frequency) routing by optical add/drop multiplexer and the action of polarization switch made of SOA. Section-3 covers the method of all-optical decimal to frequency encoded binary data generation. Method of developing frequency encoded all-optical logic units are presented in section-4. An all optical binary memory unit is presented in section-5 and conclusion is drawn in section-6.

2. Some important functions of SOA as the elements of optical processor

2.1 Frequency conversion exploiting Nonlinear Polarization Rotation (NPR) of the probe beam in SOA

One of the important properties of SOA is non- linear polarization rotation of the probe beam due to optically induced nonlinear refractive index in a bulk SOA by highly intense pump beams [Guo L.Q., Connelly M.J.(2005),(2006),(2007); Dutta N.K. et.al.,2006, Liu Y., et.al.,2003, Fu S. et.al.,2007]. During the interaction of the intense pump beam with probe beam in nonlinear SOA, the intense pump beam can modify the optical properties of the SOA which, in turn modify the intensity of probe beam as well as its SOP. If a linearly polarized light is coupled in a SOA, after leaving the SOA its SOP will change. A polarization beam splitter (PBS) at the output end can detect the nonlinear polarization rotation in terms of intensity difference. The mechanism is explained below.

At first the SOA is to be biased with suitable current and also the power level of input pump beams 'A' and 'B' are to be adjusted properly. 'X' is the linearly polarized probe beam of frequency ν but weak in intensity and, it is coupled with the pump beams in SOA. The scheme is shown in Fig.1(a).

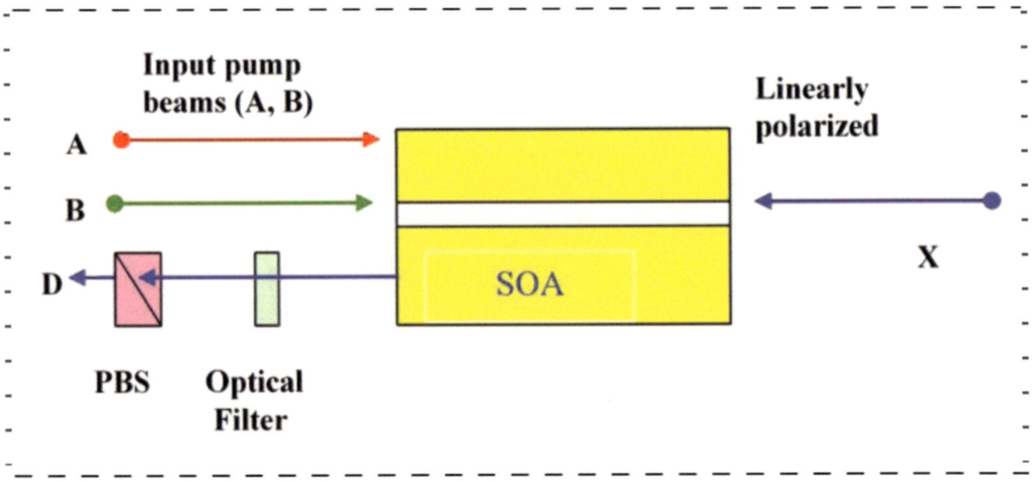

Fig. 1(a). Frequency conversion by SOP of probe beam

In the absence of both the input pump beams 'A' and 'B', the polarizer is adjusted in such a way that the pass axis of the polarization beam splitter (PBS) is crossed with respect to SOP of the linearly polarized probe beam(X). For this setting no light is obtained at output end (D). One input pump beam (A/B) alone does not change the SOP of the probe beam dramatically and no light will pass through the PBS. When both the input pump beams 'A', 'B' are present, the state of polarization of the probe beam will change drastically and as result a considerable amount of light of frequency v will pass through the PBS and will appear the output end 'D'. It is to be noted that only one pump beam of intensity equal to the sum of the intensity of both the pump beams A and B can also rotate the state of polarization of the probe beam in SOA and therefore with the help of the control beam (pump beam) of such intensity it is possible to transmit the probe beam from input end to output end of an SOA.

2.2 Function of an add/drop multiplexer

Input optical data signals may be of different frequencies and these data signals can be directed through separate paths using add/drop multiplexer (ADM) [Yu S., et.al.,2005; Jiang Y., et.al,2010]. The function of an 'ADM' is to separate a particular channel without interference from adjacent channels. This can be achieved by using an integrated 'SOA' with a tunable filter with it. The filter can be tuned at different frequencies by changing the bias current of SOA. The selected channel is reflected by the filter, amplified a second time by the Multi Quantum Well (MQW) section and extracted to a drop port by means of a circulator. The remaining channels pass through the filter section without any drop. In Fig.1(b) the frequency 'v_2' is extracted from incoming signals of frequencies v_1, v_2,v_3 ..... v_N using drop multiplexer and also added to the next step by means of add multiplexer using another circulator.

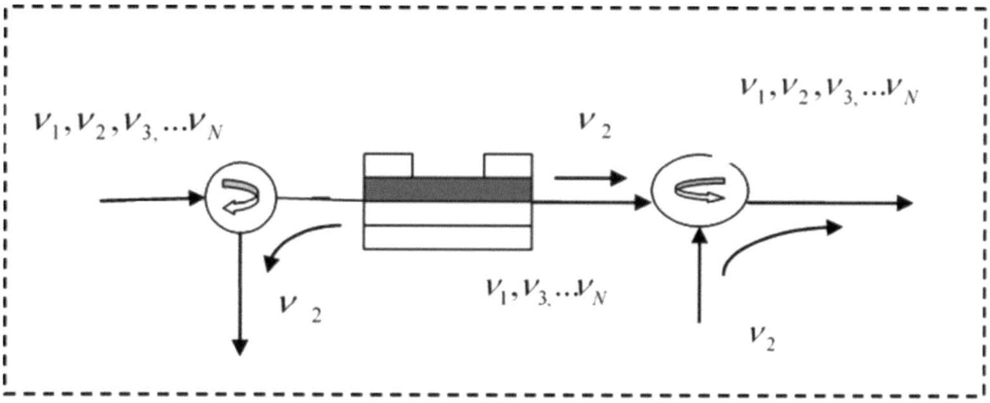

Fig. 1(b). SOA based add/drop multiplexer

2.3 Action of polarization-switch (PSW)

The principle of polarization SOA–gain saturation property may be exploited to design the nonlinear polarization switching (PSW) [Dorren H.J.S.,et-al.,2003; Garai S.K., Mukhopadhyay S.(2010); Garai S.K.,2011a]. The scheme of the polarization switching is shown in Fig.1(c). It is consisting of two laser sources having different frequencies, three polarization controllers, one strained bulk SOA, one polarization beam splitter (PBS), an attenuator and a power meter. The probe beam is a CW laser of frequency ν_1 whereas the pump laser beam is a highly intense beam of frequency ν_2. The state of polarization of the probe beam, pump beam and output beam of SOA are controlled by polarization controllers PC1, PC2 and PC3 respectively. The probe beam is fed to one input terminal of SOA via an attenuator so that the input probe beam power injected to the SOA be very low (-15 dB_m) and it confirms the operation of SOA in the linear regime under the action of probe beam alone. The orientation of linearly polarized probe beam is adjusted by PC1 in such a way that the polarization direction of the input probe beam be approximately 45^0 to the orientation of SOA layer. The output beam of SOA is combined by means of polarization beam splitter (PBS). The PBS is used to split the SOA output into horizontal (H) and vertical polarization component (V). The vertical component of SOA output is obtained at port-1 and horizontal component at port-2.

In the absence of pump beam, the optical field of linearly polarized probe beam may be decomposed into a transverse electric field (TE) and transverse magnetic field (TM) components. These two modes propagate through SOA independently and amplify by the biasing current in SOA. The biasing current is set to such a value (162 mA) that the maximum gain is obtained for TE and TM modes which are almost equal. Under this situation the state of polarization of output beam of SOA is oriented in such a way by PC3 that the beam at the output port-1 becomes zero (it is measured by power meter) i.e. vertical component (V) of the output beam of SOA is absent and obviously maximum power is delivered at port-2.

SOA have the property of polarization dependent gain saturation. Therefore, in the presence of highly intense pump beam the polarization dependent gain saturation character give rise to different refractive index change for TE and TM. Under gain saturation condition the output of port-2 will be a function of saturation-induced phase difference between two modes [Dorren H.J.S.,et-al.,2003] given by

$$\varphi = \varphi^{TE} - \varphi^{TM} = \frac{1}{2}\left[\frac{\alpha^{TE}\Gamma^{TE}g^{TE}}{v_g^{TE}} - \frac{\alpha^{TM}\Gamma^{TM}g^{TM}}{v_g^{TM}}\right]L \tag{3}$$

where L is the length of SOA, v_g^{TE} is the group velocity of the envelop of the optical electric field for TE mode, Γ^{TE} is the confinement factor, g^{TE} is the real gain function , α^{TE} is the phase modulation parameter and α_{int}^{TE} is the modal loss. All the parameters corresponding to superscript TM and TE represent the parameters for TM mode and TE mode of propagation respectively. At the PBS, the two modes coherently combine. If the phase difference φ is an odd multiple of π , the angle of rotation of the beam after combination of TE and TM mode (having almost same amplitude) at output end of SOA is $\theta = \pi / 2$ and then at the output port-2 no beam will appear. In this case the output from port-2 will be suppressed i.e. switched off. Here the induced phase difference π is controlled by the power of input pump beam as well as choosing the suitable parameters and length of SOA (intensity > 0.4 mW) [Garai S.K.,2010,2011a].

Thus in the absence of pump beam, probe beam will appear at port-2 (ON-state) and in the presence of the pump beam of specific intensity, the probe beam will be suppressed in port-2(OFF-state). Obviously the state of port-1 will be complementary with respect to port-2 i.e., in the presence of the pump beam, power will develop at port-1.

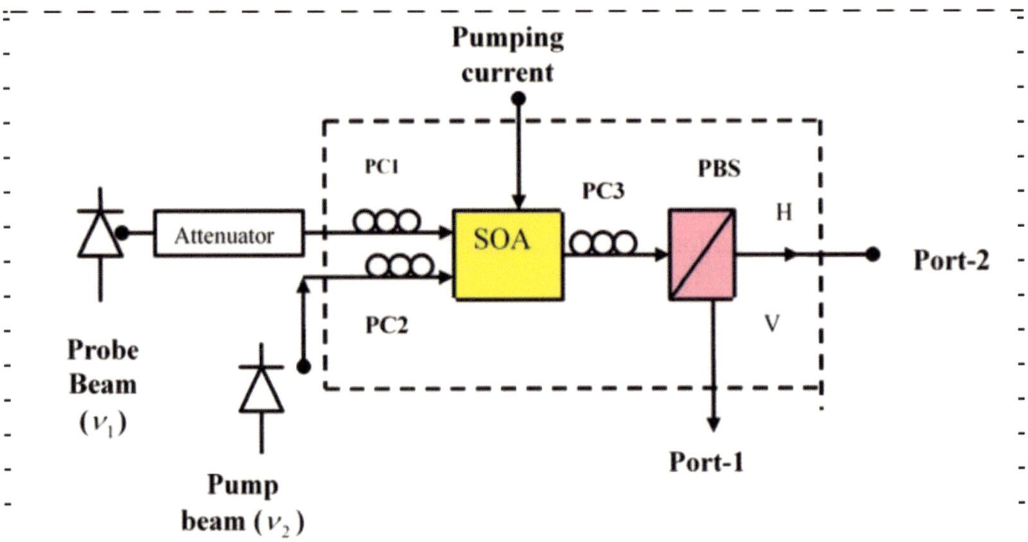

Fig. 1(c). SOA acting as a polarization switch

3. Conversion of decimal number to frequency encoded binary data

To implement the frequency encoded all-optical arithmetic logic (ALU) processors, generation of frequency encoded binary data is very important. In this section the author has first mentioned a method of generating intensity encoded binary data to frequency encoded binary data and subsequently explained the scheme of conversion of decimal data to frequency encoded binary data using the above mentioned action of PSW made of SOA.

The scheme of conversion of all optical decimal data to frequency encoded binary data works based on the principle of frequency conversion by polarization switches (PSW) and it is explained with the help of Fig.2(a) [Garai S.K.,2010,2011a]. The optical circuit comprises two polarization switches PSW1 and PSW2. Major part of the output beam of PSW1 is applied as the input pump beam of PSW2 and the rest part is coupled with the output beam of PSW2. The probe beam X1 of PSW1 is of frequency v_1 and the probe beam of PSW2 is X2 of frequency v_2.

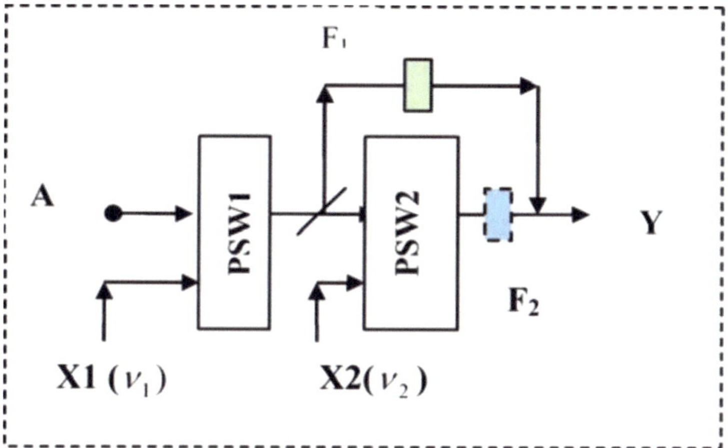

Fig. 2(a). Optical circuit for converting decimal to frequency encoded binary data

Decimal Number	Binary Number in terms of	
	'0' and '1'	'v_1' and 'v_2'
0	0000	v_1 v_1 v_1 v_1
1	0001	v_1 v_1 v_1 v_2
2	0010	v_1 v_1 v_2 v_1
3	0011	v_1 v_1 v_2 v_2
4	0100	v_1 v_2 v_1 v_1
5	0101	v_1 v_2 v_1 v_2
6	0110	v_1 v_2 v_2 v_1
7	0111	v_1 v_2 v_2 v_2
8	1000	v_2 v_1 v_1 v_1
9	1001	v_2 v_1 v_1 v_2
10	1010	v_2 v_1 v_2 v_1
11	1011	v_2 v_1 v_2 v_2
12	1100	v_2 v_2 v_1 v_1
13	1101	v_2 v_2 v_1 v_2
14	1110	v_2 v_2 v_2 v_1
15	1111	v_2 v_2 v_2 v_2

Table 1. Decimal numbers and their corresponding frequency encoded binary numbers

In the absence of input pump beam 'A', the PSW1 will be in ON state which in turn will suppress PSW2. The least fraction of the output beam of PSW1 of frequency v_1 will appear at the output. In the presence of the input beam A, the PSW1 will be in OFF state which in turn will switch the PSW2 in ON state and thereby the beam of frequency ν_2 will be obtained at the output end.

The above mentioned technique has been exploited for the conversion of decimal (0 to 15) to binary data and it is explained with the help of Fig.2(b).

Fig.2(b) comprises four frequency converter units made of polarization switches (PSW_0, $PSW_0/$), (PSW_1, $PSW_1/$), (PSW_2, $PSW_2/$) and (PSW_3, $PSW_3/$). PSW_0, PSW_1, PSW_2 and PSW_3 have their common probe beam 'X_1' of frequency ν_1, whereas another four prime polarization switches ($PSW_0/$ to $PSW_3/$) have their common probe beam 'X_2' of frequency ν_2. D_0, D_1, D_2.....D_{15} are sixteen input terminals corresponding to decimal numbers 0,1,2,3,….,15 respectively, through which optical beam of specific power is to be applied to convert a specific decimal number(corresponding to terminal number) into its binary form. For example, to convert the decimal number '9' to its binary form, a laser source of specific power [Garai S.K.,2011a] is to be applied in the terminal D_9 by means of an optical switch. The beam after entering via the terminal D_9 will split up into two equal parts and serve as the pump beam of PSW_3 and PSW_1. The beam entering via the terminal D_{13} will serve as the pump beam of PSW_3, PSW_2 and PSW_0, the beam entering via the terminal D_{15} will act as the pump beam for all four polarization switches PSW_3, PSW_2, PSW_1 and PSW_0 and so on. The splitting of the beams after entering through the sixteen terminals (0 to 15) and their function as the pump beam for different PSWs are presented in Table-2. The terminal D_0 has no internal connection to any of the polarization switches. The output ends of the combination of polarization switch (PSW_3, $PSW_3/$), (PSW_2, $PSW_2/$), (PSW_1, $PSW_1/$) and (PSW_0, $PSW_0/$) are designated as Y_3,Y_2,Y_1 and Y_0 respectively and these will give the frequency encoded

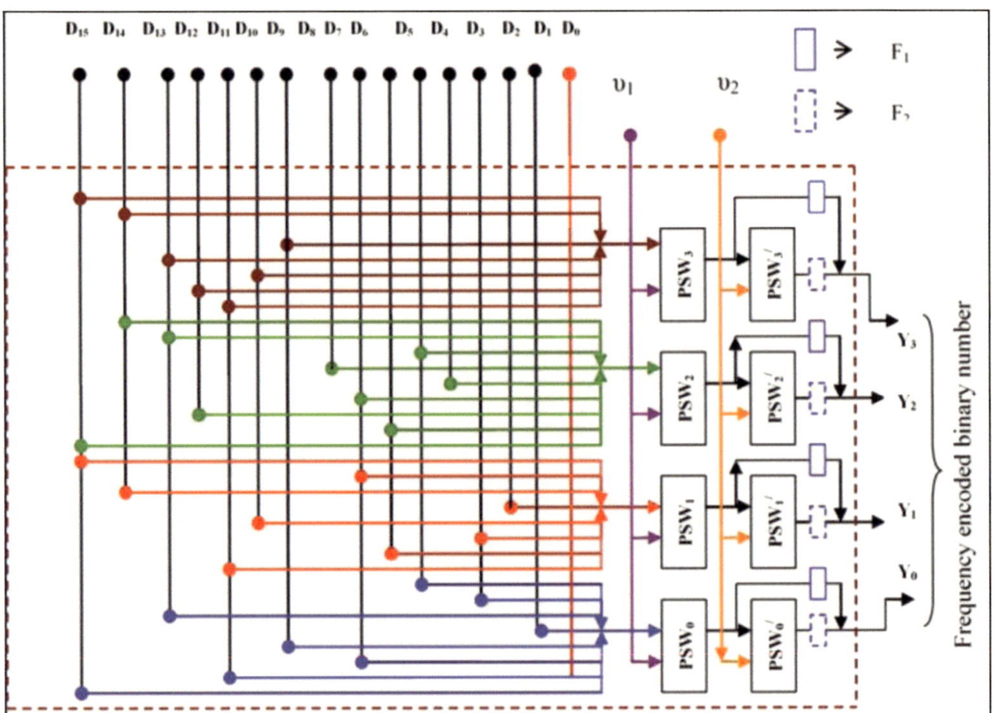

Fig. 2(b). Decimal to frequency encoded binary data conversion scheme

binary number in sequence '$Y_3Y_2Y_1Y_0$' corresponding to the input decimal number. Here Y_3 represents the most significant bit (MSB) and Y_0 represents the least significant bit (LSB) of the converted binary number.

Decimal Number	Optical beam connecting terminal	No of split up parts of beam	Connected to PSW switches as pump beam	PSW in		PSW/ in		Output
				ON state	OFF state	ON state	OFF state	Y_3 Y_2 Y_1 Y_0
0	D_0	NIL	None	All	None	None	All	$\nu_1\nu_1\nu_1\nu_1$
1	D_1	No	PSW0	1,2,3	0	0	1,2,3	$\nu_1\nu_1\nu_1\nu_2$
2	D_2	No	PSW1	0,2,3	1	1	0,2,3	$\nu_1\nu_1\nu_2\nu_1$
3	D_3	2	PSW0 PSW1	2,3	0,1	0,1	2,3	$\nu_1\nu_1\nu_2\nu_2$
4	D_4	No	PSW3	0,1,2	3	3	0,1,2	$\nu_1\nu_2\nu_1\nu_1$
5	D_5	2	PSW0 PSW3	1,2	0,3	0,3	1,2	$\nu_1\nu_2\nu_1\nu_2$
6	D_6	2	PSW1 PSW3	0,2	1,3	1,3	0,2	$\nu_1\nu_2\nu_2\nu_1$
7	D_7	3	PSW0 PSW1 PSW3	2	0,1.3	0,1,3	2	$\nu_1\nu_2\nu_2\nu_2$
8	D_8	No	PSW3	0,1,2	3	3	0,1.2	$\nu_2\nu_1\nu_1\nu_1$
9	D_9	2	PSW0 PSW3	1,2	0,3	0,3	1,2	$\nu_2\nu_1\nu_1\nu_2$
10	D_{10}	2	PSW1 PSW3	0,2	1,3	1,3	0,2	$\nu_2\nu_1\nu_2\nu_1$
11	D_{11}	3	PSW0 PSW1 PSW3	2	0,1.3	0,1,3	2	$\nu_2\nu_1\nu_2\nu_2$
12	D_{12}	2	PSW2 PSW3	0,1	2,3	2,3	0,1	$\nu_2\nu_2\nu_1\nu_1$
13	D_{13}	3	PSW0 PSW2 PSW3	1	0,2,3	0,2,3	1	$\nu_2\nu_2\nu_1\nu_2$
14	D_{14}	3	PSW1 PSW2 PSW3	0	1,2,3	2,2,3	0	$\nu_2\nu_2\nu_2\nu_1$
15	D_{15}	4	PSW0 PSW1 PSW2 PSW3	None	All	All	None	$\nu_2\nu_2\nu_2\nu_2$

Table 2. Decimal to binary conversion scheme in tabular form

Now the mode of conversion of the decimal number '0' and '13' into its frequency encoded binary number are explained with the help of Fig. 3(b).
To convert the decimal number '0' to its binary form, the laser beam is to be connected to the input terminal D_0. As the terminal D_0 has no internal connection to any of the polarization switch, therefore, polarization switches PSW_0, PSW_1, PSW_2 and PSW_3 will not get any pump beam. All these switches will get only the probe beam of frequency ν_1 from common source X_1 and therefore, all these switches will remain in ON state and the amplified probe beam of frequency ν_1 will appear at the output end of each polarization switch. Now all the polarization switches $PSW_0/$ to PSW3/ will get the pump beam from previous PSWs as well as the probe beams of frequency ν_2 from common supply X_2. Combination of the pump beam and the probe beam will drive all the polarization switches ($PSW_0/$ to $PSW_3/$) to OFF state. Fractional parts of the output beam of PSWs of frequency ν_1 after passing through bypass path of PSW/s will appear at output end of $PSW_0/$ to $PSW_3/$.
Hence at the output end, one will obtain the binary form of frequency encoded data '$\nu_1 \nu_1 \nu_1 \nu_1$', for input decimal number '0'.

To convert decimal number '13' into its binary form, the laser beam is to be connected to X_{13} terminal. After entering through D_{13}, it will split up into three equal parts. Here the three successive spilt up parts will act as pump beam for PSW_3, PSW_2 and PSW_0 unit respectively. The pump beams in these three units will switch off the PSWs which in turn will switch on $PSW_3/$, $PSW_2/$ and $PSW_0/$ unit and one will obtain the amplified probe beam of frequency ν_2 at each of the output end Y_3, Y_2 and Y_0. Remaining PSW_1 units will not get any pump beam and according to its function, one will get optical beam of frequency ν_1 at the output terminal Y_1. Thus, the binary number corresponds to the decimal number '13' is '$\nu_2 \nu_2 \nu_1 \nu_2$'.
Similarly the conversion of all other decimal number to its binary form can be explained with the help of Fig.2(b) and Table-2.
The above mentioned scheme may be extended to convert decimal numbers to binary coded decimal numbers and gray code and vice versa exploiting the above principle and that are explained in details in the work of Garai S.K.,2011a.

4. Method of developing frequency encoded different logic operations

The author was presented a method to develop all optical frequency encoded binary logic gates such as NOT, AND, OR, NAND, NOR, EX-OR etc. based on the conjugate beam generation technique by PPLN waveguide and subsequently frequency routing by add/drop multiplexers and frequency conversion using reflecting semiconductor optical amplifiers (RSOA)[Garai S.K., Samanta D.,et.al.,.(2008), Garai S.K., Mukhopadhyay S.,2009a,2011]. Conversion efficiency of conjugate beam generation by PPLN is not high enough and considerable amount of energy is lost to implement the logic operation. This problem was undertaken by the author and he tried to avert the intermediate conjugate beam generation, as a consequence he has supplanted the method by a new one. In this section, the author has presented a novel method to design all optical frequency encoded different logic gates exploiting the principle of nonlinear rotation of the state of polarization rotation (SOP) of the probe beam in semiconductor optical amplifier in the presence of pump beam of specific intensity ranges. Here conjugate beam generation is not required. Hence the conversion efficiency and speed of operation are higher compared to the earlier method. The truth table of frequency encoded different logic gates are presented in Table-3.

Input data of frequency		Output of different logic gates					
A	B	AND	OR	NAND	NOR	X-OR	X-NOR
ν_1	ν_1	ν_1	ν_1	ν_2	ν_2	ν_1	ν_2
ν_1	ν_2	ν_1	ν_2	ν_2	ν_1	ν_2	ν_1
ν_2	ν_1	ν_1	ν_2	ν_2	ν_1	ν_2	ν_1
ν_2	ν_2	ν_2	ν_2	ν_1	ν_1	ν_1	ν_2

Table 3. Truth table of frequency encoded different logic units

The scheme of the experiment for implementing frequency encoded NAND logic operation exploiting the nonlinear rotation of the state of polarization of the probe beam is shown in Fig.3. 'A' and 'B' are two input terminals through which frequency encoded pump beams are applied. 'ADM1' and 'ADM2' are the optical add and drop multiplexers which are tuned for reflected frequency 'ν_1' by the application of proper biasing current of SOAs in 'ADMs' [Garai S.K., Mukhopadhyay S., 2009, 2009b; Garai S.K., 2011c]. The reflected signal of frequency 'ν_1' from 'ADM1' is dropped down by circulators 'C_1' and then power of the beam is divided into two equal parts by means of 'beam splitter'(BS) .One part of the beam is injected as the pump beam for 'SOA1' and another part is injected as pump beam for 'SOA2'. The reflected signal of frequency 'ν_1' from 'ADM 2' is dropped down by circulator 'C_2' and then the beam is divided into two equal parts by means of beam splitter(BS). One of the beams is injected as the pump beam of SOA1 and another part is injected as pump beam for SOA3. The destination of the input beam 'A' of frequency ν_2 as the pump beam after passing through ADM1 is given by { SOA3, SOA4} and that of the input beam 'B' of frequency ν_2 as the pump beam after passing through ADM2 is given by { SOA2, SOA4}. X_1 and X_2 are two linearly polarized input probe beams of frequency 'ν_1' and 'ν_2' respectively. The state of polarizations are maintained by polarization controllers(PC).The beam X_2 is split up into three equal parts which are serving as the weak probe beam of SOA1, SOA2 and SOA3 respectively. Output of each 'SOA' is selected by an optical filter each having pass frequency equal to its corresponding input probe beam frequency. The final output is 'Y' which is obtained by connecting the output of each SOA after passing through polarization beam splitters (PBS). Initially the state of polarization of input probe beams are oriented in such a way that output from each PBS is zero in the absence of pump beams.
Now the NAND logic operation is explained with the help of Fig3.

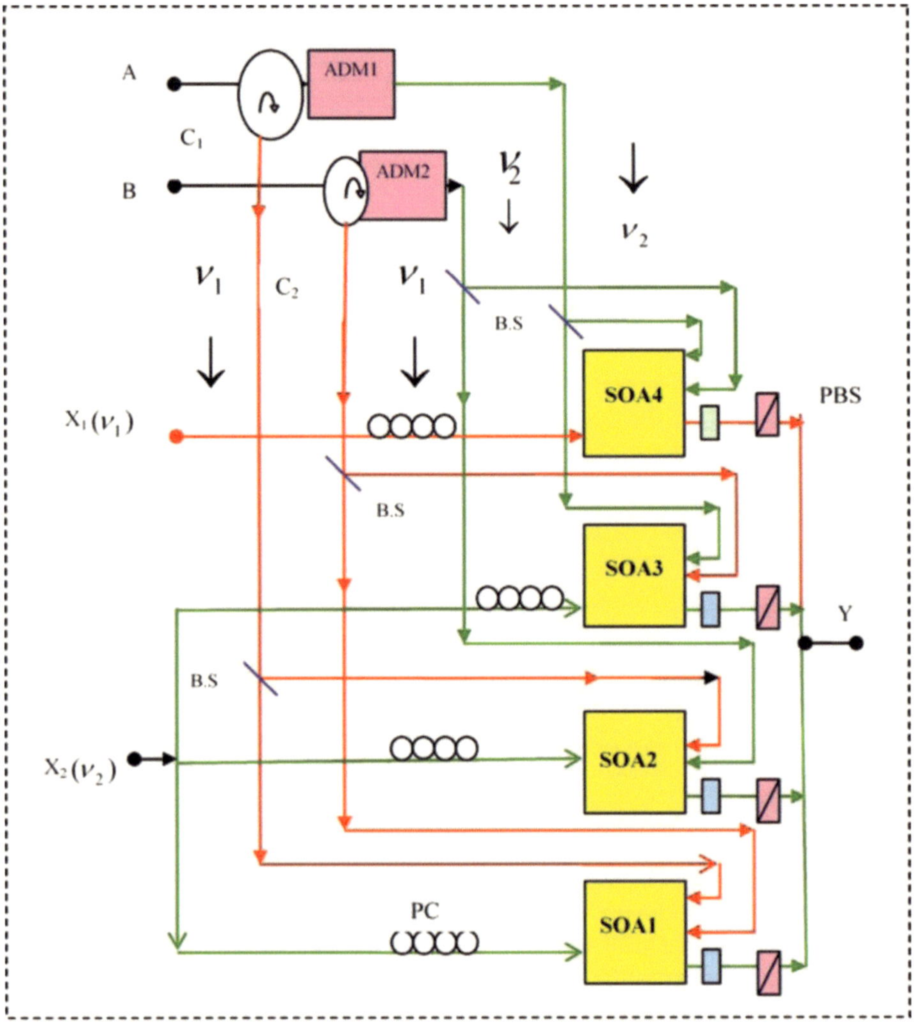

Fig. 3. Scheme of implementing frequency encoded NAND logic operation

Case-1: Both the input pump beams A and B are of frequency 'ν_1' i.e. both are at '0' state

Now both the pump beams of frequency 'ν_1' will be reflected from 'ADM1' and 'ADM2' and dropped down by circulators 'C_1 and C_2 respectively'. The destination of the input beam 'A' of frequency ν_1 as the pump beam is given by { SOA1, SOA2} and that of the input beam 'B' of frequency ν_1 as the pump beam is given by { SOA1, SOA3}.Thus SOA1 will get both the input pump beams whereas all other 'SOAs' get at most one pump beam at a time. Therefore both the pump beams of SOA1 can significantly rotate the state of polarization of input probe beam 'X_2'of frequency ν_2 and an polarization beam splitter(PBS) at the output end can detect the nonlinear polarization rotation in terms of intensity difference. As a result output beams of 'SOA1' will give a beam of frequency ν_2 at the cost of the input pump beam each of frequency ν_1.

Case-2: Input pump beam 'A' is of frequency 'ν_1' i.e. at '0' state and the B is of frequency 'ν_2' i.e. at '1' state

Now the destination of the input beam 'A' of frequency ν_1 as the pump beam after reflecting back by ADM1 is given by { SOA1, SOA2} and that of the input beam 'B' of frequency ν_2 as the pump beam after passing through ADM2 is given by { SOA2, SOA4}. Under this situation 'SOA2' only will get both the pump beams at the same time. These two pump beams can significantly rotate the state of polarization of input probe beam 'X_2' of frequency ν_2 and as a result output beam of 'SOA2' will give a beam of frequency ν_2 .

Case-3: Input pump beam 'A' is of frequency 'ν_2' i.e. at '1' state and the 'B' is of frequency 'ν_1' i.e. at '0' state

Now the destination of the input beam 'A' of frequency ν_2 as the pump beam after passing through ADM1 is given by { SOA3, SOA4} and that of the input beam 'B' of frequency ν_1 as the pump beam after reflecting back by ADM2 is given by { SOA1, SOA3}. Therefore under this situation 'SOA3' only will get both the pump beams. These pump beams can significantly rotate the state of polarization of the probe beam 'X_2' and as a result output beam of 'SOA3' will give the beam of frequency ν_2 at the output end of PBS.

Case-4: Both the pump beams are of frequency 'ν_2' i.e. both are at '1' state

Now the destination of the input beam 'A' of frequency ν_2 as the pump beam after passing through ADM1 is given by {SOA3, SOA4} and that of the input beam 'B' of frequency ν_2 as the pump beam after reflecting back by ADM2 is given by {SOA3, SOA4}. Thus both the input pump beams are injected at 'SOA4' whereas other SOAs get at most one pump beam. Therefore both the pump beams of 'SOA4' can significantly rotate the state of polarization of input probe beam 'X_1' of frequency ν_1 and as a result output beam of 'SOA4' will give a beam of frequency ν_1 at the output end.

Thus using input pump beams of frequencies ν_1 and ν_2 as input data, it is possible to get a frequency encoded NAND logic operation. NAND logic gate is the universal logic gate and all other logic gates can be developed using NAND gates only.

The utility of the above mentioned scheme is that the same circuit can be used to implement any one out of the 16 binary logic operations, only by properly selecting the frequency of the probe beam of the four SOA units. As for example, if the frequency of the probe beams SOA1 and SOA4 unit be ν_1 (X1) and that of SOA2 and SOA3 unit be ν_2 (X2), then it is possible to execute frequency encoded X-OR logic operation using the same circuit.

The block diagram of frequency encoded different logic units with proper distribution of probe beams X1(ν_1) and X2(ν_2) in four probe beam terminals of SOA units i.e., SOA1, SOA2, SOA3 and SOA4, designated by 1,2,3 and 4 respectively are as shown in Fig.4.

The above mentioned scheme may be extended to design all optical multiplexer and demultiplexer [Garai S.K., Mukhopadhyay S.(2009)], data comparator[Garai S.K.(2011)] multivalued logic unit such as trinary [Garai S.K., 2010], quaternary etc. logic gates and all optical arithmetic logic unit [Garai S.K.(2011c)].

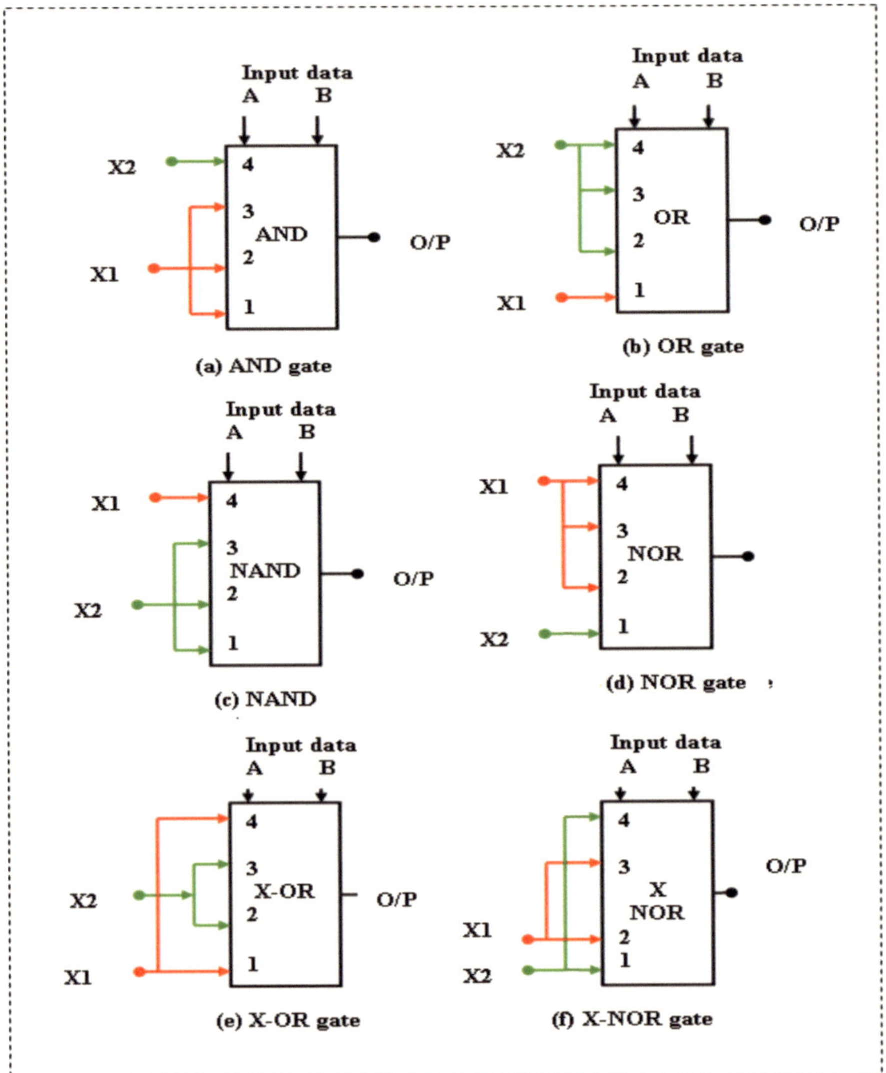

Fig. 4. Block diagram of different logic units

5. All optical frequency encoded memory unit

The very fast running optical memory and optical logic gates are the basic building blocks for any optical computing and data processing system. Realization of a very fast memory-cell in the optical domain is very challenging one. In last two decades many methods of implementing all-optical flip-flops have been proposed. Most of these suffer from speed limitation because of slow switching response of the active devices [Zhang S. et.al.,2005, Ghosal et.al.,2008, Fatehi M.T. et.al.,1984]. In this present chapter the author presents a method of developing a frequency encoded memory unit based on the polarization switching action of semiconductor optical amplifier (SOA) using frequency encoded data [Garai S.K., Mukhopadhyay S.,2010].

The basic building blocks of the memory unit consists of three polarization switches PSW1 and PSW2 [Garai S.K., 2010, 2011a], an isolator, two input sources X_2 and X_1 giving the probe beams having frequencies v_2 and v_1 respectively and one add/drop multiplexer, ADM as shown in Fig.5. The beam obtaining from output port-2 of PSW1 splits up into two parts by means of beam splitter B.S. One part of the beam is coupled as the input pump beam for polarization switch PSW2 and another part is serving as the output data (Y). Similarly, the output beam from port-2 of PSW2 is split up into two parts. One part is serving as the probe beam of PSW3 switch via an attenuator A_T and another part is viewed as output at Y terminal via the attenuator. The low intensity input probe beam of PSW3 switch is controlled by the isolator. The function of the isolator is that it allows the part of the output beam of PSW2 to appear at Y end but prevents the output of PSW1 to appear at the input end of PSW3. 'A' is the input pump beam terminal of switch PSW1. The input pump beam is injected to PSW1 via add/drop multiplexer ADM. The ADM is tuned for reflection frequency v_2. The reflected beam of frequency v_2 is reflected back by ADM and drop down by circulator and injected as the pump beam (control beam) for PSW3. The beam obtained at the output port-2 of PSW3 is coupled with input pump beam 'A' by a beam coupler (B.C.).

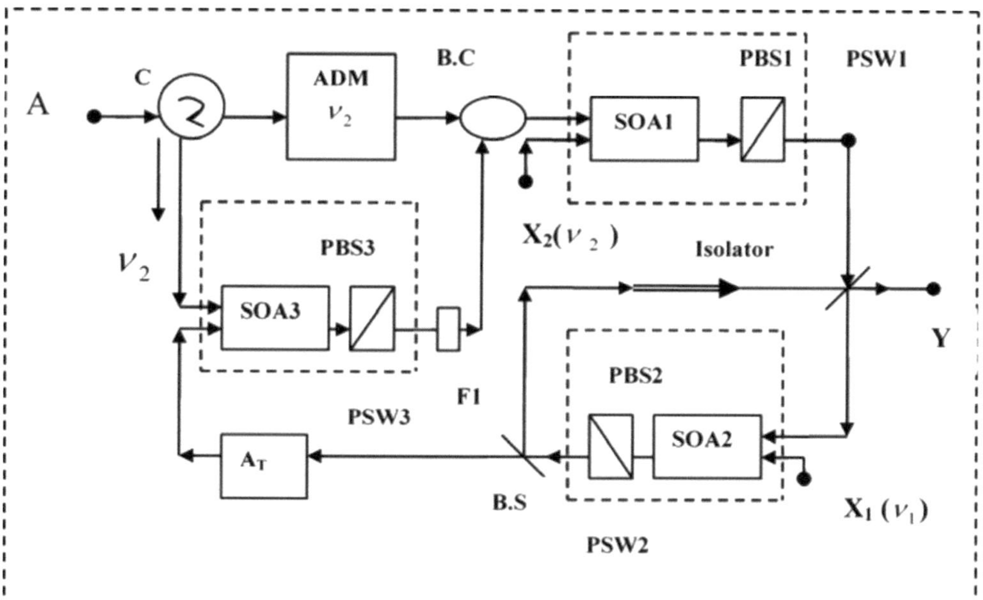

Fig. 5. Frequency encoded single bit memory circuit

The operation of the frequency encoded memory unit is now explained with the help of Fig.6. Here the frequency of optical signal ' ν_1 (corresponding wavelength λ_1) is encoded as 0 state and the frequency ν_2 (corresponding wavelength λ_2) as the state 1.
If the input beam 'A' be of frequency $v_1(0)$, then it will pass through ADM and behaves as the pump beam for polarization switch PSW1. As the probe beam X_2 of PSW1 is of frequency v_2, therefore, by the joint action of pump and probe beam the PSW1 goes to switch off state i.e. output of PSW1 will give no signal (zero). Now the polarization switch PSW2 will get only probe beam signal X_1 of frequency v_1 and according to the action of polarization switch

the PSW2 will be in ON state i.e. output of PSW2 is the signal of frequency v_1. A fraction of the beam of frequency v_1 will be displayed at the final output and the intensity of the remaining part will be attenuated to a value so that a desired low intensity beam is serving as probe beam of PSW3 switch. In PSW3 since no pump beam is present, the probe beam of frequency v_1 will appear at the output port-2 and finally it is coupled with the input pump beam of PSW1 of frequency v_1. Therefore optical beam of frequency $v_1(0)$ will remain at the output end (Y) of the memory unit.

Now if the input beam A of frequency v_1 is removed from the circuit, the output beam of frequency v_1 of port-2 of PSW3 will serve as the input pump beam for switch PSW1 which leads to switch off the PSW1 and in turn it will switch on PSW2. Thus the signal of frequency $v_1(0)$ will continue to remain at the output end Y.

If the input beam 'A' is of frequency $v_2(1)$ then it will be reflected back by ADM and drop down by circulator and behave as the pump beam for PSW3. As no pump beam for switch PSW1 is present, so this switch will come to the ON state and the amplified probe beam X_1 of frequency v_2 will appear at the output end of PSW1.This output beam with the joint action of probe beam X_2 switch off the PSW2. Therefore no signal will be obtained from output end of PSW2. Now no signal probe beam being present at PSW3, no probe beam will appear at the output port-2. Again probable leakage pump beam of frequency v_2 in port-2 is blocked by v_1 pass filter F_1. Therefore no beam from the output end of PSW3 will be injected as pump beam for switch PSW1. Thus PSW1 will remain at ON state giving constant output signal of frequency v_2 when the input signal A is of frequency v_2.

Input beam A of frequency	Data stored at output (Y)
v_1	v_1
OFF	v_1
v_2	v_2
OFF	v_2
v_1/v_2	v_1/v_2
OFF	Last input data

Table 4. Excitation table of frequency encoded memory unit.

Now if the input signal 'A' of frequency v_2 is removed, both the pump beam and probe beam will be absent at the input end of PSW3 and as a result no output beam will appear at the port-2. Again no pump beam being present at the input end of PSW1 switch, it will remain in on state giving amplified probe beam of frequency v_2 at the output end. This output beam in turn will drive the switch PSW2 to OFF state and no beam will be obtained at the output port-2 of PSW3. Thus when the input beam of frequency v_2 is withdrawn, the

signal of 'frequency- $v_2(1)$' will remain stored at the output end 'Y'. The excitation table of the memory unit is as shown in Table-4.
This scheme may be extended to design multivalued memory unit with some extra circuit elements [Garai S.K., Mukhopadhyay S.(2010)] as well as designing multivalued flip-flops [Garai S.K.,2012]exploiting the same working principle.

6. Conclusion

Whole operation is all-optical one, so one can expect a very high speed of operation from the system. Considering the present scenario of speed and band width limitation of electronic computing, signal processing and future problem of data traffic, the author has developed all these frequency encoded all optical logic units, and memory unit which will be very useful in all optical computing and the optical networking. All these optical gates and memory units are suitable to perform so many advanced functions in communication based network such as in all-optical bit pattern recognition, all-optical bit-error rate monitoring, all optical packet addressing and pay-load separation, all optical label swapping, all optical packet drops in optical time domain multiplexing (OTDM) etc. The frequency encoded all these all optical logic processors are expected to be very useful in present days as well as in near future for wavelength division multiplexing and demultiplexing networks.
Here the author has selected the wavelength of the encoded inputs signals corresponding to the encoded signal of frequencies $\nu_1(0)$ and $\nu_2(1)$ in C band (1536 nm -1570 nm) and these are respectively 1540 and 1550nm. The advantages of using C-Band is that here the frequency conversion gain is almost independent of frequency. The separation between two consecutive encoded wavelength '5 nm' is sufficient. The function of optical 'add/drop multiplexer' is very specific about frequency of reflection and it merely allow to pass a spreading of frequency. Again, at the output end as only the beam of one frequency is obtained at a time, therefore, there is no question of crosstalk. To maintain the state of polarization (SOP) of probe beams polarization controller (necessary polarizer) is to be used. The performance of SOA based optical logic processors are preferred as SOA based optical switches are more efficient because of its higher nonlinearity with least switching power (<-3dB_m) and high switching contrast ratio (20dB). Here the speed of the operation is depending on the switching speed of SOA based state of polarization rotation of the probe beam as well as the switching speed of coupled version of different circuit elements within the interconnecting fibers. It also depends on the distances of different units and propagation distance between two SOAs. The operating speed of SOA switch is restricted to 100 Gb/s due to its response time of gain saturation in regular SOA. Though switching speed of individual circuit element is very high (of the order of sub Pico second), however, the speed of the couple version will be reduced to 40 to 50 GHz due to propagation delay (order of nanosecond) within the interconnecting fibers. However very fast response (100 Gb/s) can be achieved using quantum dot SOA-MZI switch [Ju H.,et.al.,2005; Sun H., et.al.,2005; Vyrsokinos K., et.al.,2010] and quantum dot SOA as polarization rotation switches with an integrated circuit. The fast switching action of SOA enhances the speed of logic operation and as a result the speed of processing becomes faster for multi-bit operation. Therefore the above-mentioned scheme demands for overall feasibility, practicality and versatility of designing all optical logic processor system with very high speed.

7. References

Asghari M., White I.H., Penty R.V., 'Wavelength conversion using semiconductor optical amplifiers, J.Lightwave Technology, Vol.15, No.17,1997, pp.1181-1190

Awwal A.A.S., Karim M.A.(1990), 'Microprocessor design using polarization encoded optical shadow casting', Appl. Opt., Vol.29, No.14,1990, pp. 2107-2112.

Chakraborty B., Mukhopadhyay S.(2009), 'Alternative approach of conducting Phase-modulated all-optical logic gates'', Optical Engineering,Vol. 48, No.3, March 2009, pp.035201-5.

Connelly M.J.(2008), 'Semiconductor Optical Amplifiers', Kluwer Academic Publishers, 2002

Dorren H.J.S., Lenstra D., Liu Y, Hill M. T., Khoe G.D.(2003), 'Nonlinear Polarization Rotation in Semiconductor Optical Amplifiers: Theory and Application to All-Optical Flip-FlopMemories", IEEE Journal of Quantum Electronics, Vol.39, No.1, 2003, pp.141-148.

Dutta N.K., Wang Q.(2006), 'Semiconductor Optical Amplifiers', World Scientific Publishing, Singapore, 2006(Chapter 8).

Fatehi M.T., Wasmundt K.C., Collins S.A.(1984), 'Optical flip-flops and sequential logic circuits using a liquid crystal light valve', Appl. Optics, Vol.23, No.13, 1984,pp.2163-2171.

Fu S., Zhong W.D., Shum P., Wu C., Zhou J.Q.(2007), 'Nonlinear Polarization rotation in semiconductor optical amplifiers with linear polarization maintenance', IEEE Photonics TechnologyLetters,Vol.19, No.23,2007, pp.1931-1933.

Garai S.K.(2010), 'A scheme of developing frequency encoded tristate-optical logic operations using Semiconductor Optical Amplifier', Journal of Modern Optics,Vol.57, No.6,2010, pp.419-428.

Garai S.K.(2011), 'A method of developing frequency encoded multi-bit optical data comparator using Semiconductor Optical Amplifier', Optics and Laser Technology, Vol.43, No.1, 2011, pp.124-131.

Garai S.K.(2011a), 'Method of all-optical frequency encoded decimal to binary and BCD, binary to gray' and gray to binary data conversion using semiconductor optical amplifiers', Applied Optics, Vol.50, No.21,2011,pp.3795-3807.

Garai S.K.(2011b), 'A novel method of designing all optical frequency encoded Fredkin and Toffoli logic gates using semiconductor optical amplifiers', IET Optoelectronics, Vol.5, No.6, 2011, pp.247-254.

Garai S.K.(2011c), "A novel all-optical frequency encoded method to develop Arithmetic and Logic Unit (ALU) using semiconductor optical amplifiers," IEEE/OSA Journal of Light wave Technology, Vol.29, No.23, 2011, pp. 3506-3514.

Garai S. K.(2012), 'A novel method of developing all optical trinary JK, D-type and T-type flip-flops using semiconductor optical amplifiers', Applied Optics, (In Press-20.12.2011)

Garai S. K., Mukhopadhyay S.(2009), 'Method of implementing frequency encoded multiplexer and demultiplexer systems using nonlinear Semiconductor Optical Amplifiers', Optics and Laser Technology, ,Vol.41, No.8,2009, pp.972-976 .

Garai S. K., Mukhopadhyay S.(2009a), 'Method of implementing frequency encoded NOT, OR and NOR logic operations using Lithium niobate waveguide & Reflecting Semiconductor Optical Amplifiers', PRAMANA-journal of physics, Vol.73, No.5, 2009, pp.901-912.

Garai S.K., Mukhopadhyay S.(2010), 'A novel method of developing all-optical frequency encoded memory unit exploiting nonlinear switching character of Semiconductor Optical Amplifier', Optics and Laser Technology,Vol.42, No.7, 2010, pp.1122-1127.

Garai S.K., Mukhopadhyay S.(2011), 'A scheme of developing frequency encoded tristate logic operations exploiting non-linear character of PPLN waveguide and RSOA', Optik,Vol.122, No. 6, 2011, pp. 498-501.

Garai S.K., Mukhopadhyay S.(2009b), 'Method of implementation of all-optical frequency encoded logic operations exploiting the propagation characters of light through Semiconductor Optical Amplifiers', J.Opt,Vol.38, No.2, 2009, pp.88-102.

Garai S.K., Samanta D., Mukhopadhyay S.(2008), 'All-optical implementation of inversion logic operation by second harmonic generation and wave mixing character of some non-linear material', Optics and Optoelectronic technology,Vol6, No.4,2008, pp.39-42.

Garai S.K., Pal A., Mukhopadhyay S.(2010), 'All optical frequency encoded inversion operation with tristate logic using Reflecting Semiconductor Optical Amplifiers' Optik, Vol.121,No.16,2010,pp.1462-65

Ghosal A. K., Basuray A.(2008), 'Trinary flip-flops using Savart plate and spatial light modulator optical computation in multivalued logic', Optoelectronics Letters, Vol.4,2008, pp. 0443-0446 .

Jiang Y., Luo X., Hu L., Wen J., Li Y.(2010), 'All optical add-drop multiplexer by utilizing a single semiconductor optical amplifier', Microwave and Optical Technology Letters,Vol.52, No.9,2010, pp.1977–1980.

Ju H., Zhang S., Lenstra D., de Waardt H., Tangdiongga E., Khoe G. D., Dorren H. J.S.(2005) 'SOA-based all-optical switch with subpicosecond full recovery', Optics Express, Vol.13, No. 3, 2005, pp.942-947

Lacey J.P.R., Summerfield M.A., Madden S.J.(1998), ' Tunability of polarization insensitive wavelength converters based on four-wave mixing in Semiconductor Optical Amplifiers, J.Lightwave Technology, Vol.16, No.12, 1998, pp.2419-2427.

Liu Y., Hill M.T., Tangdiongga E., Waardt H.de, Calabretta N., Khoe G.D., Dorren H.J.S.(2003), 'Wavelength conversion using Nonlinear Polarization Rotation in a Single Semiconductor Optical Amplifier', IEEE Photonics Technology Letters, Vol.15, No.01, 2003,pp.90-92.

Soto H., Erasme D., Guekos G.(1999), 'Cross-polarization modulation in Semiconductor Optical Amplifier', IEEE Photon. Technol. Lett. Vol.11, No.8,1999,pp.970-972.

Sun H., Wang Q., Dong H., Dutta N.K.(2005), 'XOR performance of a quantum dot semiconductor optical amplifier based Mach-Zender interferometer', Optics Express,Vol.3, No.06,2005, pp.1892-1899.

Toyohiko Y.(1986), 'Optical space-variant logic gate based on spatial encoding technique', Optics Letters,Vol.11, No.4, 1986, pp. 260-262.

Vyrsokinos K., Stampoulidis L. , Gomez-Agis F. , Voigt F. K., Zimmermann L., Wahlbrink T., Sheng Z. , Thourhout D.V. , Dorren H.J.S.,(2010) 'Ultra-high Speed, all-optical wavelength converters using single SOA and SOI photonic integrated circuits', Photonics Society Winter Topicals Meeting Series (WTM), IEEE,2010,pp.113–114, doi:10.1109/PHOTWTM.2010.5421936

Yu S., Gu W.(2005), 'A tunable wavelength conversion and wavelength add/drop scheme based on cascaded second order nonlinearity with double pass configuration', IEEE J. Quantum Electron, Vol.41. No.7, 2005, pp. 1007-1012

Zaghloul Y.A., Zaghloul A.R.M.(2006), 'Unforced polarization based optical implementation of Binary logic', Optic Express, Vol.14, No.16, 2006, pp.7252-7269.

Zaghloul Y. A., Zaghloul A. R. M. and Adibi A. (2011), 'Passive all-optical polarization switch, binary logic gates, and digital processor', Vol.19, No.21, 2011, pp.20332-20346

Zhang S., Li Z., Liu Y., Khoe G. D. and Dorren H. J. S.(2005) 'Optical shift register based on an optical flip-flop memory with a single active element', Optics Express , Vol.13,No.24,2005,pp.9708-9713

4

SOA-Based Optical Packet Switching Architectures

V. Eramo[1], E. Miucci[1], A. Cianfrani[1], A. Germoni[2] and M. Listanti[1]
[1]*DIET Sapienza University of Rome,*
[2]*Co.Ri.Tel.*
Italy

1. Introduction

The service evolution and the rapid increase in traffic levels fuel the interest toward switching paradigms enabling the fast allocation of Wavelength Division Multiplexing WDM channels in an on demand fashion with fine granularities (microsecond scales). For this reason, in the last years, different optical switching paradigms have been proposed (Sabella et al., 2000): optical-packet switching (OPS), optical-burst switching (OBS), wavelength-routed OBS, etc. Among the various all-optical switching paradigms, OPS attracts increasing attention. Owing to the high switching rate, Semiconductor Optical Amplifier (SOA) is a key technology to realize Optical Packet Switches. We propose some Optical Packet Switch (OPS) architectures and illustrate their realization in SOA technology. The effectiveness of the technology in reducing the power consumption is also analyzed. The chapter is organized in three sections. The main blocks (Switching Fabric, Wavelength Conversion stage, Synchronization stage) of an OPS are illustrated in Section 2 where we also show some examples of realizing wavelength converters and synchronizers in SOA technology. Section 3 introduces SOA-based single-stage and multi-stage switching fabrics. Finally the SOA-based OPS power consumption is investigated in Section 4.

2. Optical packet switching architectures

The considered optical switch architecture (Eramo, 2000; 2006; Sabella et al., 2000) is shown in Fig. 1. It has N input and output fibers, each fiber supports a WDM signal with M wavelengths, so an input (or output) channel is characterized by the couple (i, λ_j) wherein i $(i \in 1, \cdots, N)$ identifies the input/output fiber and λ_j, $(j \in 1, ..., M)$ identifies the wavelength. In general, optical packet switches can be divided into two categories: slotted (synchronous) and unslotted (asynchronous) networks. In a synchronous switch (Eramo, 2000), as illustrated in Fig. 1 packets with fixed length are aligned (synchronized) by synchronizers before they enter the switch fabric. This type of switch generally achieves a fairly good throughput since the behavior of the packets is regulated. However, complex and expensive synchronization hardware is needed at each node. On the other hand, in an asynchronous switch (Eramo et al., 2003), the packets are not aligned and they are switched one by one on the fly. Asynchronous networks generally have lower cost, better flexibility, and robustness, but usually they have lower overall throughput than synchronous networks. The switch architecture is equipped with a number r of WCs which are shared according to a particular strategy (Eramo et al., 2009b). At each input line, a small portion of the optical power is tapped to the electronic controller. The switch control unit detects and reads packet headers and drives the space switch matrix and the WCs. Incoming packets on each input line are wavelength demultiplexed (DEMUXs blocks in Fig. 1). An electronic control logic, on the basis of the routing information contained in each packet header, handles packet contentions and decides which packets have to be wavelength shifted. Packets not requiring wavelength conversion are directly routed towards the output lines; on the contrary, packets requiring wavelength conversions will be directed to the pool of r WCs and, after a proper wavelength conversion, they will reach the output line. An example of realization of synchronizers and wavelength converters in SOA technology is shown in Sections 2.1 and 2.2 respectively. Section 3 is devoted to illustrate both SOA-based single-stage and multi-stage switching fabrics.

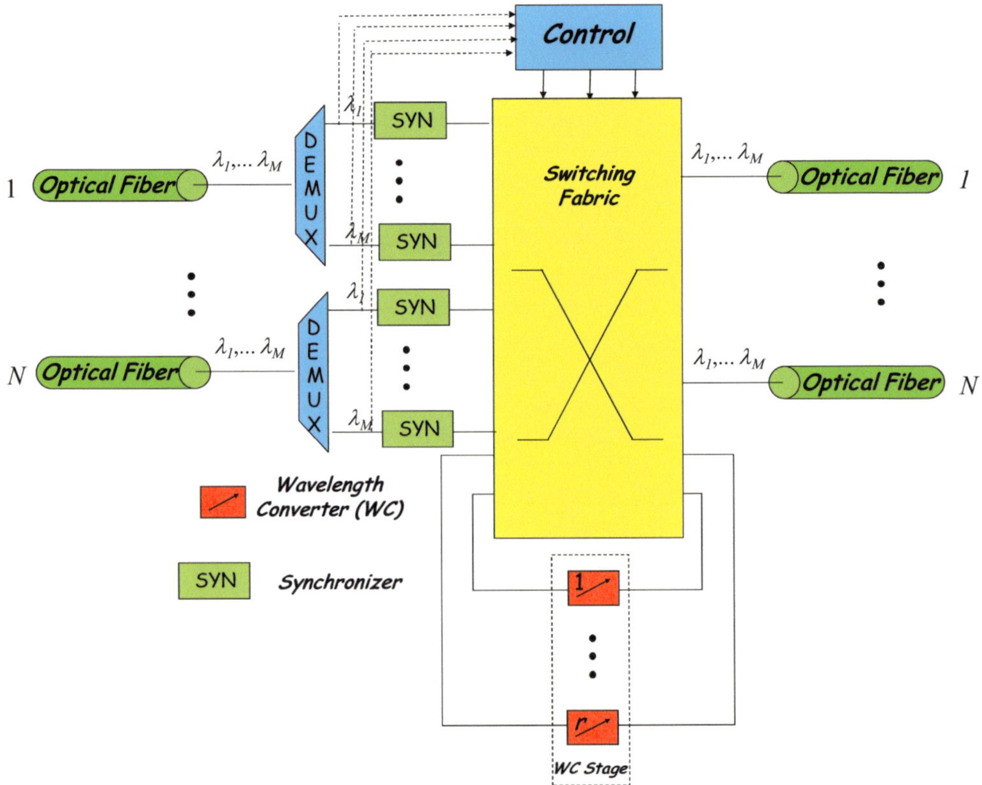

Fig. 1. Optical Packet Switching Architecture with N Input/Output Fibers, M Wavelength and r shared Wavelength Converters.

2.1 Realization of SOA-based synchronizers

The synchronizers are used in the synchronous optical packet switches where the packets have a fixed size and their arrivals on each wavelength are synchronized on a time-slot basis by synchronization devices located at the ingress of the switch before the switching operation is performed. Most of the synchronizers (Chao et al., 2000; Zucchelli et al., 1998) are composed of a series of optical switches designed to select the proper optical path and pairs of fiber delay lines with different optical lengths of $\frac{T_s}{2^k}$ (T_s:time slot; k:integer). In these architectures, however, increasing the number of switches to improve the time resolution causes additional increases in optical loss and crosstalk. To overcome loss problems SOA-based synchronizers have been proposed. Next we illustrate and explain two of them. In the first one (Sakamoto et al., 2002) synchronization is achieved by selecting one of some optical paths, each with a different length, using wavelength and space switching based on a wavelength-tunable distributed Bragg reflector laser diode (LD) and n semiconductor optical amplifier (SOA) gates per channel. The synchronizer has its own internal reference clock. The clock period equals the time slot duration (T_s) and the synchronizer aligns input packets with the time slot packet by packet. Synchronization is achieved by counting each delay of each input packet with respect to the reference time and choosing the optical paths with the appropriate length. Fig. 2 shows the schematic structure of the synchronizer. Each channel is equipped with a wavelength-divisionmultiplexing (WDM) coupler, a wavelength converter, an optical splitter, semiconductor optical amplifier (SOA) gates, two stages of fiber delay lines, an optical coupler, arrayed waveguide gratings (AWGs) for MUX/DEMUX, a delay counter, and a wavelength-tunable laser. The out-of-band optical label switching technique is used, in which optical packet and optical labels are carried on different wavelengths (Okada et al., 2001). The delay counter estimates the delay of each optical label and selects one of m wavelengths of the tunable laser and one of n SOA gates. The wavelength of each optical packet signal is converted to the laser wavelength by the wavelength converter. The wavelength-converted optical packet signal passes through one of n SOA gates and the first-stage delay lines, each of which has a different delay time of $\frac{T_s}{n}$. The packet signal then passes through one of m AWG ports and the second-stage delay lines, each with delay time difference of $\frac{T_s}{n \times m}$. Consequently, there are optical paths with different lengths and synchronization is attained.

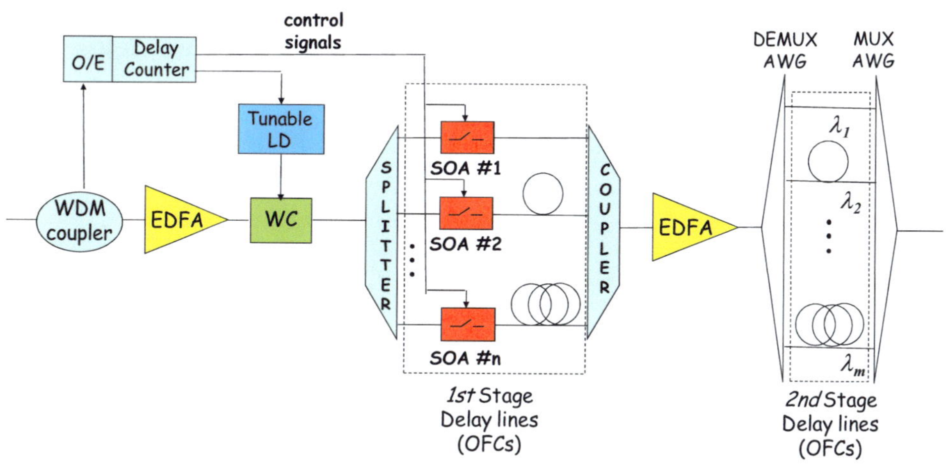

Fig. 2. Schematic structure of the synchronizer. WC: Wavelength converter. SOAG: SOA gate. OFC: Optical fiber circuit.

The second synchronizer has been proposed in (Mack et al., 2008) and it is illustrated in Fig. 3. Feed-forward structure with SOA-based gates is used here because of its high operation speed, large tuning range, and the potential for integration within the large SOA-based switch

(Mack et al., 2008). The synchronizer is composed by N_{SYN} stages. There are one 1×2 splitter, one 2×1 coupler, two SOAs, two optical bandpass filter(OBF) and one FDL in each stage. SOAs were used as the gates to select the required delay and compensate for losses. In order to suppress accumulated amplified spontaneous emission, Optical Bandpass Filters was placed inside each synchronizations stage.

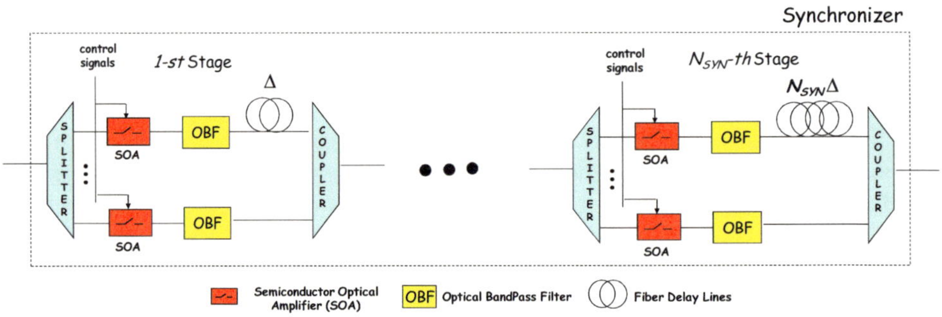

Fig. 3. Fiber based synchronizer with SOA gates and Optical Bandpass Filter (OBF). Δ is the delay introduced by the fiber delay line of the $1st$ stage.

2.2 Realization of SOA-based Wavelength Converters

In packet switching networks, tuneable wavelength converters can be used to resolve packet contention and overcome the optical buffering problem. An example of SOA-based Wavelength Converter is illustrated in Fig. 4. It is referred to as Delayed Interference Signal WCs (DISCs) and has been proposed in (Sakaguchi et al., 2007). DISC employs nonlinear effect in SOA and utilizes an SOA and an OBF placed at the amplifier output. It can be constructed by using commercially available fiber-pigtailed components. It has a simple configuration and allows photonic integration.

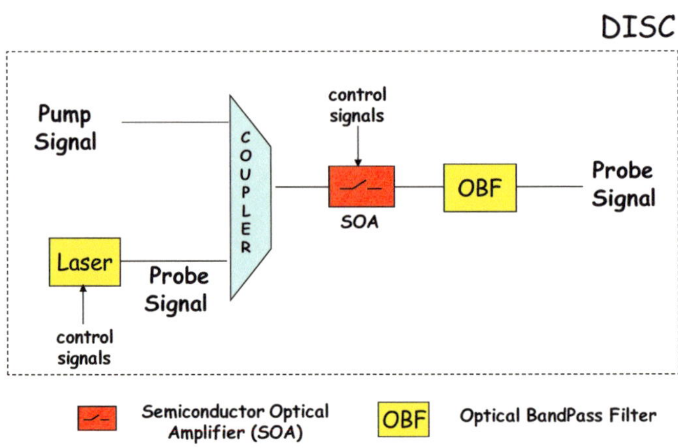

Fig. 4. Realization of an all-optical wavelength converter by using a Delayed-Interference Signal Wavelength Converter (DISC).

3. SOA-based switching fabric

Switching fabric in Future Optical Packet Switches require high-speed optical switches (or gates). That can either be optically or electrically controlled. Such optical switches can be constructed using SOAs due to their high switching rate. The simplest method to control an SOA gate is by turning the device current on or off. The great advantage of SOA gates is that they can be integrated to form gate array. Next we illustrate SOA-based Single-Stage and Multi-Stage switching in Sections 3.1 and 3.2 respectively.

3.1 Single-stage approach

The structure of a switching fabric depends on the adopted sharing strategy of the Wavelength Converters. Two of them are reported in Figs 5 and 6.

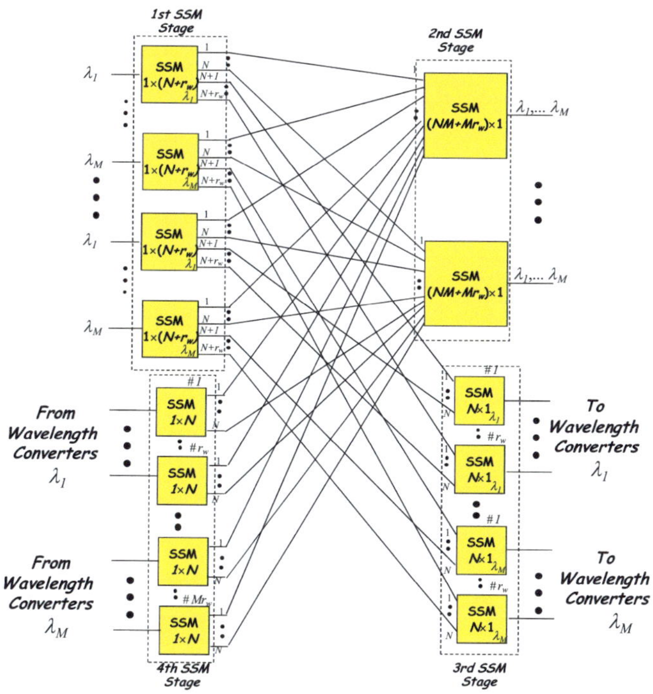

Fig. 5. Shared-Per-Wavelength Single-Stage Optical Switching Fabric.

They are related to two proposed main WC sharing strategies. In the first one, referred to as Shared-Per-Wavelength (SPW) (Eramo et al., 2008; 2009a;b), the WCs are partially shared. All of the packets arriving on a given wavelength share the same pool of converters. In the second one referred to as Shared-Per-Node (SPN) (Eramo et al., 2009c; Eramo, 2010),the WCs are fully shared and all of the arriving packets share the same pool of WCs. Next we illustrate the switching fabrics of the SPW and SPN switches.

The SPW switching fabric is illustrated in 5 and its operation mode is the following (Eramo et al., 2008; 2009a;b). A packet not requiring wavelength conversion is directly routed towards the Output Fibers (OF). On the contrary a packet needing the use of a WC will be directed to the pool of r_w WCs dedicated for the wavelength on which the packet is arriving.

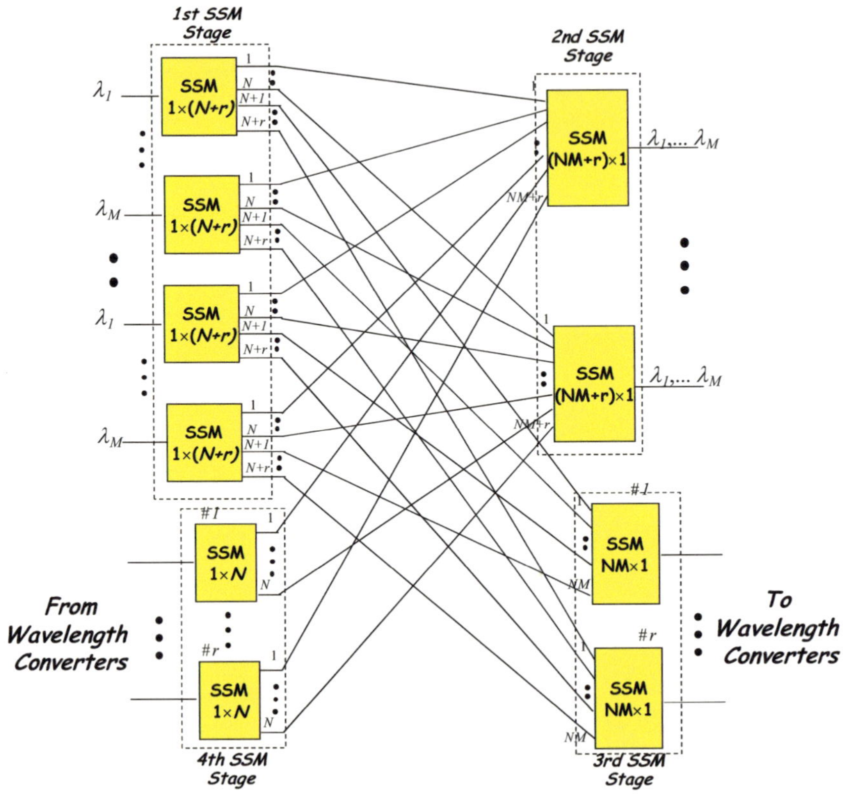

Fig. 6. Shared-Per-Node Single-Stage Optical Switching Fabric.

The selection of either an OF or a WC is realized by means of a $1 \times (N + r_w)$ Space Switching Module (SSM) of the $1st$ SSM stage. Each $N \times 1$ SSM of the $3rd$ SSM stage in Fig. 5 has the function to forward to a WC the packet selected by the control unit to be wavelength converted. After the conversion, the packets are sent to the OFs by means of a $1 \times N$ SSM of the $4th$ SSM stage. The function of an $(N + r_w)M \times 1$ SSM of the $2nd$ SSM stage is to couple all of the packets directed to any OF.

The SPW sharing strategy (Eramo et al., 2009c; Eramo, 2010) allows for a reduction in the switching fabric complexity, improving the scalability. As a matter of fact, the SSMs of the $1st$ and $3rd$ stage have reduced complexity with respect to the ones of the SPN reference switch diagrammed in Fig. 6 with r denoting the total number of shared WCs. This is due to the fact that r_w, number of WCs shared per wavelength in SPW switch, is much smaller than r, the number of WC shared in SPN switch. The reduction in switching fabric complexity of the SPW switch leads to a smaller signal attenuation and consequently to a smaller SSM power consumption.

We report in Fig. 7.a and Fig. 7.b an example of realization of $1 \times K$ SSM and $K \times 1$ SSM respectively by means of splitters, couplers and Semiconductor Optical Amplifiers (SOA). The input of an $1 \times K$ SSM is switched to the output #j by turning on SOA #j and turn off the remaining SOAs. The SOA in the $K \times 1$ SSM is activated when at least one input signal has to be coupled.

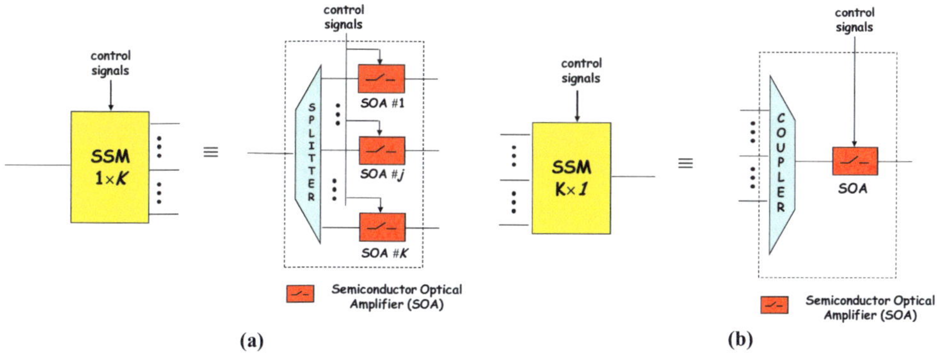

Fig. 7. Realization of $1 \times K$ SSM (a) and $K \times 1$ SSM (b) by means of splitters, couplers and Semiconductor Optical Amplifiers (SOA).

3.2 Multi-stage approach

One of the most used Multi-Stage (MS) switching fabric is the BENES one. It belongs to a class of rearrangeably non blocking networks with 2×2 switching elements. Fig. 8.a shows a 8×8 BENES switch using 20 2×2 switching elements. It is one of the most efficient architectures in terms of used number of 2×2 switching elements. A $P \times P$ BENES switch requires $\frac{P}{2}(2log_2P - 1)$ 2×2 switching elements, with P being a power of 2 (Benes, 1965). A single 2×2 switch can be realized in SOA technology as shown in Fig. 8.b. It is made by four SOAs, two splitters and two couplers and enables connectivity in both the *bar* and *crossed* states similar to a directional coupler fabricated in lithium niobate.

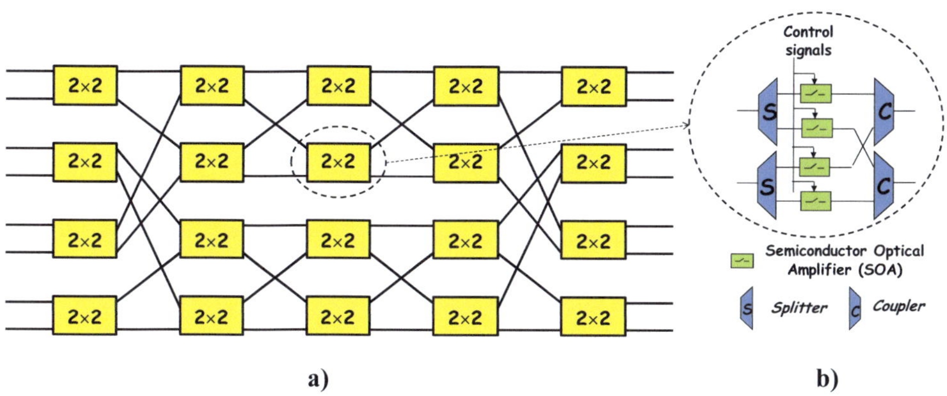

Fig. 8. 8×8 BENES switching architecture realized with 20 2×2 switching elements (a). SOA technology based 2×2 switching element (b).

A switching fabric supporting N IF/OF, M wavelengths and fully shared r wavelength converters can be realized with an $2NM \times 2NM$ BENES network. An example of BENES switching fabric is illustrated in Fig. 9 in the case N=2, M=2 and r=2.

The total number of splitters and couplers can be reduced as illustrated in the switch of Fig. 10. It is obtained by starting from the switch reported in Fig. 9 and by combining in the

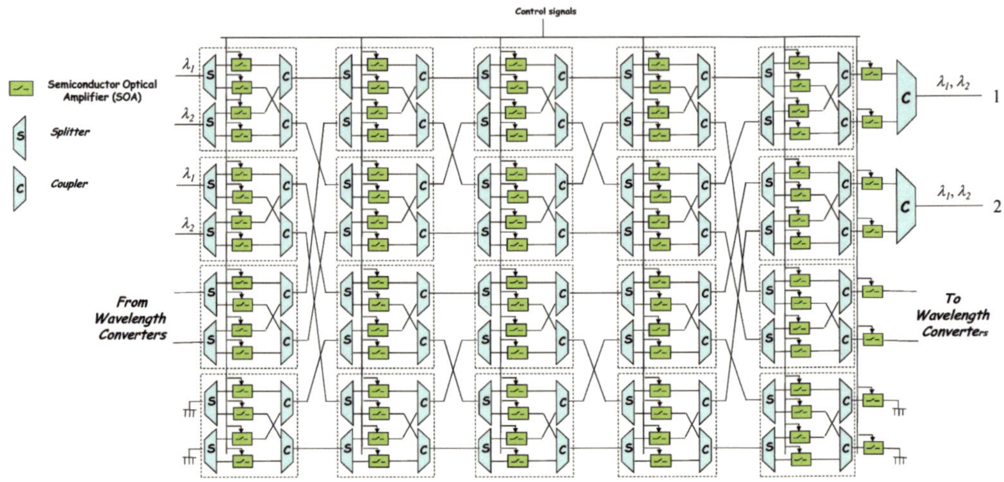

Fig. 9. BENES Optical Packet Switch realized with splitters, couplers and SOAs (N=2, M=2, r=2).

adjacent stages with a 3dB Directional Coupler (DC) the output couplers on the left-hand and the input splitters on the right-hand.

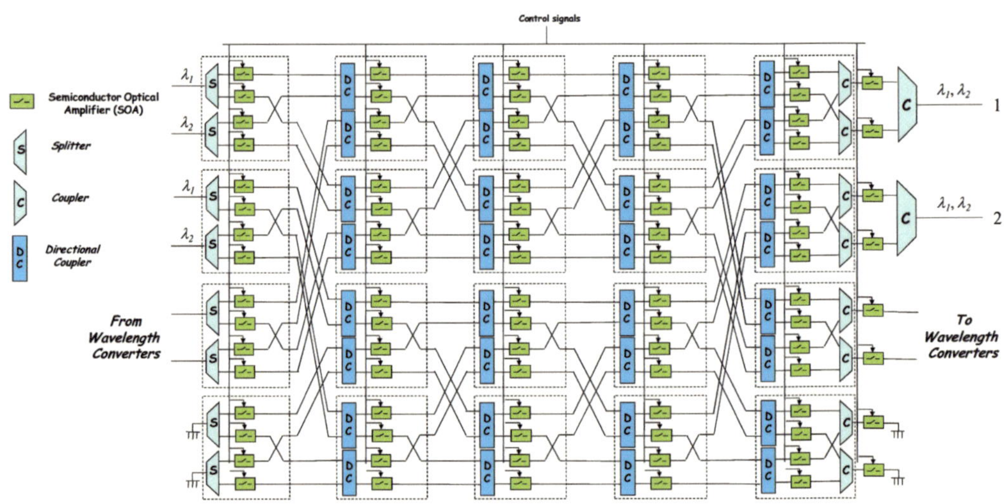

Fig. 10. BENES Optical Packet Switch realized with splitters, directional couplers, couplers and SOAs (N=4, M=2, r=2).

4. Power consumption in SOA-based optical packet switches

High capacity routers system designer are facing with power consumption issues. Today commercial products that can follow the increase in capacity demand for packet switched networks are based on multirack equipment. Optical packet switching (OPS) (Ben Yoo, 2011) systems could lead to solve this issue providing a solution that could be compact, fast, and power efficient. Next we propose some models to investigate the power consumption of Optical Packet Switching. First of all we introduce a SOA's power consumption model in Section 4.1 able to evaluate the power consumption as a function of the main SOA's parameters (current, forward polarization voltage, material loss, gain,$\cdots$). Analytical models are introduced in Section 4.2 to evaluate the power consumption of Synchronous SPN (SSPN) Optical Packet Switches equipped with SS and MS switching fabric. Similar models have been introduced for the asynchronous case (Eramo et al., 2009c; Eramo, 2010) and when the SPW sharing strategy is adopted (Akar et al., 2011; Eramo et al., 2011). Some numerical results reporting the power consumption of Optical Packet Switches are illustrated in Section 4.3.

4.1 SOA's power consumption model

The SOA's power consumption model illustrated in (Hinton et al., 2008) is adopted; the SOA's power consumption equals the supply power $P_{SOA}^{al,G}$ of the SOA needed to provide the gain G. $P_{SOA}^{al,G}$ can be expressed as follows:

$$P_{SOA}^{al,G} = V_b i_b = V_b \left(1 + \frac{lnG}{\Gamma_{SOA}\alpha_{SOA}L_{SOA}}\right) i_t \tag{1}$$

where V_b is the SOA forward bias voltage, i_b is the polarization current, Γ_{SOA} is the confinement factor, α_{SOA} is the material loss, L_{SOA} is the length and i_t is the transparency current given by:

$$i_t = \frac{q w_{SOA} d_{SOA} L_{SOA} N_0}{\tau} \tag{2}$$

where w_{SOA} is the SOA active region effective width, d_{SOA} is the active region depth, $q = 1.6 \times 10^{-9} C$ is the electronic charge, N_0 is the conduction band carrier density required for transparency, τ is the carrier spontaneous decay lifetime.

4.2 Analytical models

The analytical evaluation of the OPS power consumption is carried out as a function of the main switch and traffic parameters (Eramo, 2010; Eramo et al., 2011). We propose two analytical models to evaluate the power consumption of synchronous Optical Packet Switches equipped with Single-Stage and BENES switching fabric in Sections 4.2.1 and 4.2.2 respectively.

4.2.1 Analytical evaluation of the power consumption in SSPN OPS equipped with single-stage fabric switching

In evaluating the various power consumption in the SS-SSPN Optical Packet Switch we notice from Figs 1,6 that at time t:

- there are as many turned on synchronizers in the synchronization stage as the number $N_a(t)$ of packets forwarded;
- there are as many turned on SOAs in $1st$ stage as the number $N_a(t)$ of packets forwarded;
- the number of turned on SOAs in both $2nd$ stage and $3rd$ stage equals the number $N_c(t)$ of converted packets;
- there are as many active turned on SOAs in $4th$ SSM stage as the number $N_d(t)$ of OFs in which at least one packet is directed;

- all of the r Wavelength Converters are turned on; this assumption is a consequence of the limited speed of each WC that makes no feasible the use of a WC when only a wavelength conversion has to be performed.

According to these remarks we can write the following expression for the average power consumption $P_{av,T}^{SS-SSPN}$ for the SS-SSPN switch:

$$P_{av,T}^{SS-SSPN} = E[N_a]C^{SYN} + E[N_a]C_1^{SOA} + E[N_c](C_3^{SOA} + C_4^{SOA}) + E[N_d]C_2^{SOA} + \\ + rC_{WC} + E[N_{SOA}^{SS-SSPN,off}]C_{off}^{SOA} \quad (3)$$

wherein:

- C^{SYN} is the power consumption of a turned on synchronizer;
- C_i^{SOA} (i=1,...,4) is the power consumption of a turned on SOA in the ith stage (i=1,...,4); from Eq. 1 obviously we have $C_i^{SOA} = P_{SOA}^{al,G_i}$ (i=1,...,4) where $G_1 = N + r$, $G_2 = NM + r$, $G_3 = NM$ and $G_4 = N$ are the gains needed to overcome the loss for the turned on SOA located in the ith stage (i=1,...,4);
- C_{WC} is the power consumption of a Wavelength Converter;
- C_{off}^{SOA} is the power consumption of a turned off SOA; it is equal to $V_b i_{off}$ where i_{off} is the polarization current of an inactive SOA and needed to guarantee a high SOA switching rate (Eramo et al., 2011);
- $E[N_a]$, $E[N_d]$ and $E[N_c]$ are the steady-state average values of the random processes $N_a(t)$, $N_d(t)$ and $N_c(t)$ respectively at an arbitrary epoch; the evaluation of $E[N_a]$, $E[N_d]$ and $E[N_c]$ is carried out in in Appendix-A (Eramo et al., 2008; 2009a;c; 2011);
- $E[N_{SOA}^{SS-SSPN,off}]$ is the number of turned off SOAs; it is given by the total number $N_{SOA}^{SS-SSPN,tot} = N(N + r)M + r + Nr + N$ of SOAs to the total number $N_{SOA}^{SS-SSPN,on} = E[N_a] + 2E[N_c] + E[N_d]$ of turned on SOAs that is:

$$E[N_{SOA}^{SS-SSPN,off}] = N(N + r)M + r + Nr + N - (E[N_a] + 2E[N_c] + E[N_d]) \quad (4)$$

4.2.2 Analytical evaluation of the power consumption in SSPN OPS equipped with BENES fabric switching

Each 1×2 splitter, 2×2 directional coupler and 2×1 coupler shown in Fig. 10 introduce an attenuation of 2 that is recovered by the SOAs located after each splitter, directional coupler and coupler. If a packet is directly forwarded it goes through the BENES switch once. Conversely if the packet has to be wavelength converted the BENES switch is crossed twice and a wavelength converter is used. In particular notice as a directly forwarded packet needs the use of one 1×2 splitter, one 2×1 coupler, $2log_2 2NM - 2$ directional couplers and $2log_2 2NM$ SOAs each having a gain equal to 2. On the contrary a wavelength converted packet needs the use of two 1×2 splitters, two 2×1 couplers, $4log_2 2NM - 4$ directional couplers and $4log_2 2NM$ SOAs. Let us denote with C_{df}^{SOA} and C_{wc}^{SOA} the sum of the power consumption of the SOAs involved in the switch paths in the case in which a packet is directly forwarded and wavelength converted respectively. We can write the following expression for the average power consumption $P_{av,T}^{B-SSPN}$ of a SSPN switch equipped with BENES switching fabric:

$$P_{av,T}^{B-SSPN} = E[N_a]C^{SYN} + E[N_a]C_{df}^{SOA} + E[N_c]C_{wc}^{SOA} + rC_{WC} + E[N_{SOA}^{B-SSPN,off}]C_{off}^{SOA} \quad (5)$$

where $E[N_{SOA}^{B-SSPN,off}]$ is the number of turned off SOAs; it is given by the total number $N_{SOA}^{B-SSPN,tot} = 4NMlog_2 2NM$ of SOAs to the total number $N_{SOA}^{B-SSPN,on} = 2(E[N_a] + E[N_c])log_2 2NM$ of turned on SOAs that is:

$$E[N_{SOA}^{B-SSPN,off}] = (4NM - 2(E[N_a] + E[N_c]))log_2 2NM \quad (6)$$

Because the power consumption of turned on SOA in the BENES switching fabric equals $P_{SOA}^{al,2}$, we can simply write the following expression for C_{df}^{SOA} and C_{wc}^{SOA}:

$$C_{df}^{SOA} = 2P_{SOA}^{al,2} log_2 2NM \quad (7)$$

$$C_{wc}^{SOA} = 4P_{SOA}^{al,2} log_2 2NM \quad (8)$$

Finally notice as by inserting Eqs (6)-(8) in Eq. (5) and by using the expressions of $E[N_a]$, $E[N_d]$ and $E[N_d]$ evaluated in Appendix-A (Eramo et al., 2008; 2009a;c; 2011), we can able to calculate the average power consumption $P_{av,T}^{B-SSPN}$ of the synchronous SPN switch equipped with BENES switching fabric.

4.3 Evaluation of power consumption

We compare some Optical Packet Switching architecture by taking into account as reference the average energy consumption per bit $E_{av,T} = \frac{P_{av,T}}{NMB}$ where B denotes the bit rate carried out on each wavelength.

We perform the analysis under the following assumptions:

- the synchronizer described in Fig. 3 is used. Because in each stage and at each time only one of two SOAs is active, assuming a 3 dB attenuation for the couplers and splitters and neglecting the loss occurring in both the OBF and the short FDLs, we have the following expression for the synchronizer's power consumption:

$$P_{SYN} = N_{SYN} P_{SOA}^{al,G} |_{G=4} \quad (9)$$

- The SOA's power consumption model illustrated in Section 4.1 is adopted and allowing us, according to Eq. (1), to express the SOA power consumption as a function of the main SOA parameters (V_b, i_b, w_{SOA}, ...); A♯2 commercial SOAs Eramo (2010) produced by manufacture A is used to implement the switching fabric. The A♯2 parameter values are reported in Table 1.
- As Wavelength Converter, the Delayed Interference Signal Wavelength Converter (DISC) illustrated in Section 2.2 is used. Its power consumption has been evaluated in (Sakaguchi et al., 2007) when commercial SOA produced by some manufactures are employed. In particular we consider the B♯1 SOA characterized by a Multiple Quantum Well (MQW) type structure and produced by manufacture B. We report in Table 2 the main parameter values for B♯1. The power consumption, measured in (Sakaguchi et al., 2007), is also reported. It equals 187mW when the WC is operating at bit-rate B=40 Gb/s.

Symbol	Explanation	Value
V_b	Forward Bias Voltage	2 V
Γ_{SOA}	Confinement Factor	0.15
α_{SOA}	Material Loss	10^4
L_{SOA}	Length	$700\mu m$
w_{SOA}	Active Region Effective Width	$2\mu m$
d_{SOA}	Active Region Depth	$0.1\mu m$
N_0	Conduction Band Carrier Density	$10^{24}m^{-3}$
τ	Carrier Spontaneous Decay Lifetime	$10^{-9}s$
P_{sat}	Saturation Power	50 mW

Table 1. Main parameter values for the $A\sharp2$ commercial SOAs (Sakaguchi et al., 2007)

	Type	Active region Length (μm)	Active region width (μm)	Active region thickness (μm)	Confinement Factor	Consumed power (mW) (40Gb/s)
B#1	MQW	1100	1,25	0,038	0,2	187

Table 2. Main parameter values for the $B\sharp1$ commercial SOA (Sakaguchi et al., 2007); the power consumption of DISCs realized with $B\sharp1$ SOAs is also reported at bit-rate B=40 Gb/s.

Next, we compare the average energy consumption per bit $E_{av,T}$ of four optical packet switches (OPS) equipped with Single-Stage switching fabric: the Asynchronous Shared-Per-Wavelength (SS-ASPW) and the Asynchronous Shared-Per-Node (SS-ASPN) OPS where the WCs are per wavelength and fully shared respectively, the Synchronous Shared-Per-Wavelength (SS-SSPW) and Synchronous Shared-Per-Node (SS-SSPN) OPSs where the packets are synchronously switched and the WCs are shared per wavelength and per node, respectively. To evaluate power consumption in SS-SSPN OPS, we use the model described in Section 4.2.1. The models described in (Eramo et al., 2009c; Eramo, 2010; Eramo et al., 2011) are used to evaluate the power consumption in SS-ASPN, SS-SSPW and SS-ASPW optical packet switches. Sample switch design is reported in Fig. 11 with target Packet Loss Probability (PLP) smaller than or equal to 10^{-6}. Fig. 11 has been obtained for SS-ASPW, SS-ASPN, SS-SSPN, and SS-SSPW switches by the application of the related models (Eramo et al., 2008; 2009c; Eramo, 2010; Eramo et al., 2011) , for switch size $N = 16$, varying the offered traffic p. The number of wavelengths needed to obtain the asymptotic target PLP value (10^{-6}) is calculated first for each value of the offered load. This number of wavelengths depends on output contention only and therefore is influenced by the choice of operational context (synchronous or asynchronous) and not by the switch architecture. As a consequence, the synchronous solutions require fewer wavelengths to achieve the same PLP target. Then the minimum number of wavelength converters to reach that asymptotic PLP target value (Eramo, 2000) is determined. From Fig. 11 you can notice for a given value of offered traffic, the Shared-Per-Node switch needs fewer WCs than Shared-Per-Wavelength Node. This is obviously due to the full WC sharing strategy adopted in SPN nodes (Eramo et al., 2009b).

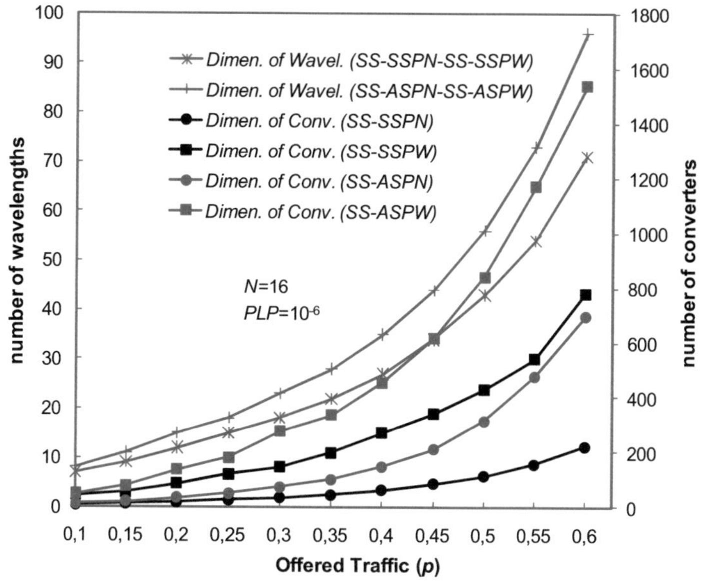

Fig. 11. Dimensioning of wavelengths and WCs in SS-ASPW, SS-ASPN, SS-SSPW and SS-SSPN switches so that the PLP is smaller than or equal to 10^{-6}. The number N of Input/Output Fibers equal 16 and the offered traffic p is varying from 0,1 to 0,6.

In Fig. 12, the average energy consumption per bit $E_{av,T}$ of switch configurations obtained in Fig. 11 as a function of the offered load is presented. Synchronizers with N_{SYN}=4 stages are considered. A bandwidth B=40Gb/s is occupied by each signal. The SS-ASPW architecture presents itself as the most power-efficient solution among all compared solutions as a consequence of the combination of asynchronous operation and wavelength converter sharing solution that allow the use of smaller Space Switching Module in the 1*st* and 3*rd* stages that leads to both smaller attenuation and SOA power consumption.

The comparison in power consumption for Synchronous SPN Optical Packet Switches equipped with Single-Stage and BENES switching fabric is reported in Fig. 13 where the average energy consumption per bit $E_{av,T}^{SS-SSPN}$ and $E_{av,T}^{B-SSPN}$ are reported versus the number N of Input/Output Fibers for M=64, N_{SYN}=16 and B=40Gb/s.

The turned off SOAs are polarized with injection current i_{off}=7mA needed to increase the switching rate. In fact the rise-time and the fall-time decrease with increasing injection current because of the strong dependence of the carrier lifetime on the carrier density (Ehrhardt et al., 1993). From Fig. 13 we can notice that for N greater than or equal to 36, $E_{av,T}^{B-SSPN}$ overcomes $E_{av,T}^{SS-SSPN}$ and the BENES switch is more efficient in energy consumption than Single-Stage switch for N increasing. That is a consequence of the linear dependence $O(Nlog_2N)$ of the number of SOA in BENES switch against the quadratic dependence $O(N^2)$ in Single-Stage switch when N increases. This different type of dependence allows a reduction in number of turned off SOAs in BENES switch with respect to the Single-Stage switch. That is confirmed in Fig. 14 where we report the number of turned off SOA versus N in Single-Stage and BENES switches for the same parameters of Fig. 13.

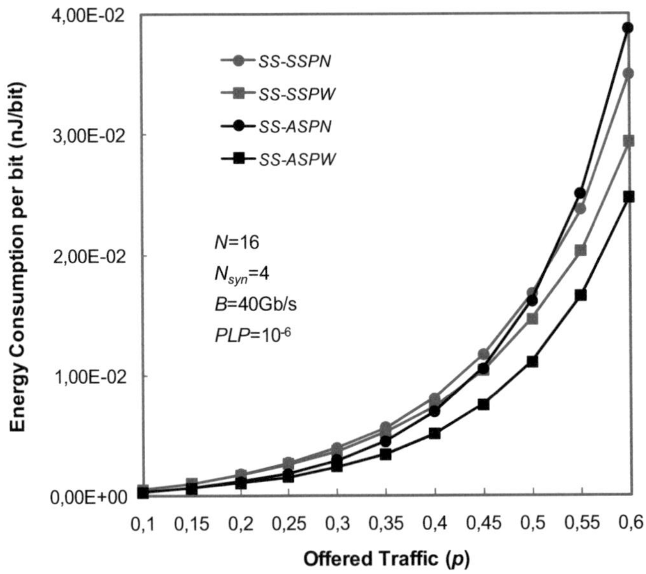

Fig. 12. Average energy consumption per bit in Single Stage(SS) ASPW, ASPN, SSPW and SSPN switches versus the offered traffic for N=16 and N_{SYN}=4. The number M of wavelengths and the number of WCs are dimensioned so that the the PLP is smaller than or equal to 10^{-6}.

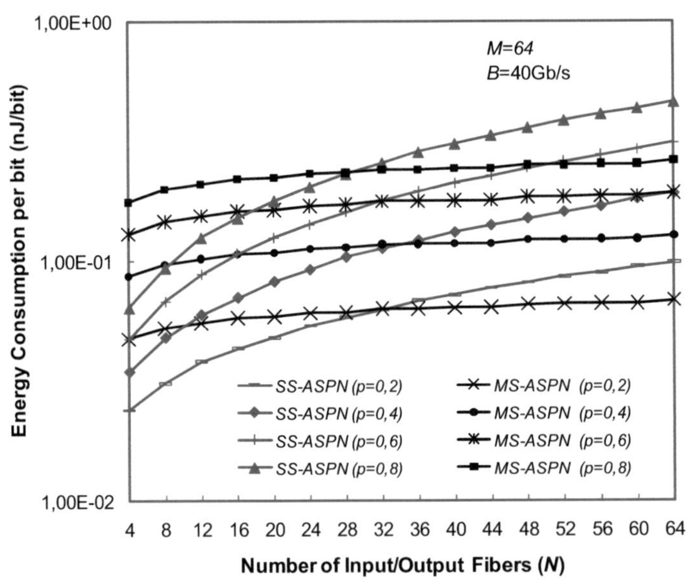

Fig. 13. Comparison of average energy consumption per bit in Single Stage (SS) SSPN and Benes (B) SSPN switches versus the number N of Input/Output Fibers for M=6 and p varying from 0,2 to 0,8. The turned off SOAs are polarized with a current i_{off}=7mA.

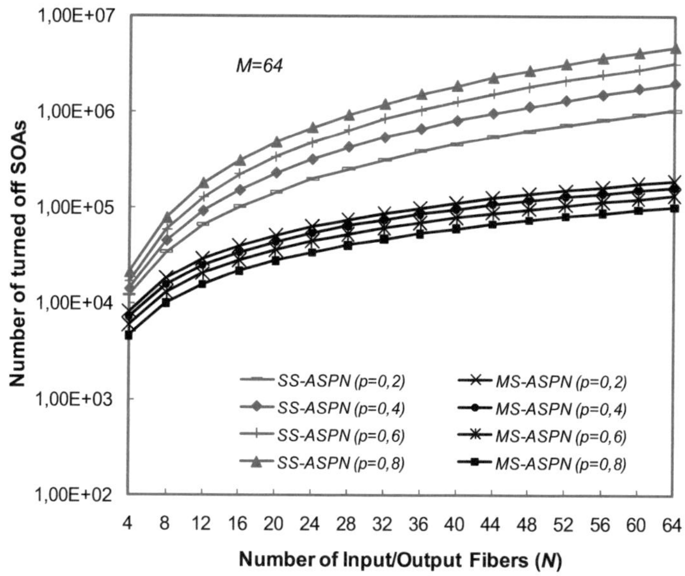

Fig. 14. Number of turned off SOAs in SS-SSPN and B-SSPN switches versus the number N of Input/Output Fibers for M=64. The offered traffic p is varied from 0,2 to 0,8.

5. Conclusions

The chapter discussed issues concerning power consumption of future high-capacity optical packet nodes. When using optical buffers, due to attenuation problems, optical nodes consumes more power than electronic nodes. For this reason we have taken into account bufferless OPS equipped with shared Wavelength Converters to solve output packet contentions. We have proposed some sophisticated analytical models in order to evaluate and compare the power consumption in OPSs equipped with Single-Stage and Multi-Stage switching Fabric. The obtained results show that in the case of OPS equipped with Single-Stage switching fabric, the combination of the asynchronous operation with the wavelength-based system partitioning in Asynchronous Shared-Per-Wavelength OPS leads to significant power saving with respect to the other solutions in the range of interest for switching fabric dimensioning. Finally we have also shown that for larger switches, the BENES switch has an energy consumption lower than the one of an Single-Stage switch if the current injection of the turned off SOAs is considered.

6. Appendix-A: Evaluation of $E[N_a]$, $E[N_d]$ and $E[N_c]$ in switches equipped with single-stage and BENES switching fabric

The evaluation of $E[N_a]$, $E[N_d]$ and $E[N_c]$ in synchronous Optical Packet Switches is carried under the following assumptions:

- packet arrivals on the $N \times M$ input wavelength channels at each time-slot are not dependent on each other;
- packet arrivals occur with probability p on each input wavelength channel;

- the destination of a packet is uniformly distributed over all N OFs, i.e., the probability that an arriving packet is directed to a given OF is equal to $\frac{1}{N}$.

Due to the synchronous operation mode of the SSPN switches, we can evaluate $E[N_a]$, $E[N_d]$ and $E[N_c]$ at a given time-slot.

The average number $E[N_a]$ of forwarded packets can be evaluated by taking into account that the packet loss can be due to either the lack of output wavelength channels or the lack of wavelength converters. We can write:

$$E[N_a] = E[N_o] - E[N_{p,wl}] - E[N_{p,cl}] \tag{10}$$

wherein:

- $E[N_o] = pNM$ is the average number of packets offered to the switch;
- $E[N_{p,wl}]$ is the average number of lost packets due to the lack of output wavelength channels. It is simply given by:

$$E[N_{p,wl}] = N \sum_{j=M+1}^{NM} (j-M) \binom{NM}{j} \left(\frac{p}{N}\right)^j \left(1-\frac{p}{N}\right)^{NM-j}. \tag{11}$$

- $E[N_{p,cl}]$ is the average number of lost packets due to the lack of WCs. The evaluation of this term has been evaluated in (Eramo et al., 2002; 2009c) by solving an urn problem (Eramo et al., 2002).

The average number $E[N_d]$ of OFs in which at least one packet is directed can be simply expressed as:

$$E[N_d] = N\left(1-\left(1-\frac{p}{N}\right)^{NM}\right) \tag{12}$$

Finally the average number $E[N_c]$ of packets forwarded with wavelength conversion can be computed by subtracting the number $E[N_d]$ of packets forwarded without wavelength conversion to the average number $E[N_a]$ of forwarded packets, that is:

$$E[N_c] = E[N_a] - E[N_d] \tag{13}$$

Finally by inserting (10)-(13) in (3) and (5) we are able to evaluate the average energy consumption $E_{av,T}^{SS-SSPN}$ and $E_{av,T}^{B-SSPN}$ of the switches equipped with Single-Stage and BENES switching fabric respectively.

7. Acknowledgment

The research leading to these results has received funding from the European Community's Seventh Framework Programme FP7/2007-2013 under grant agreements nř 247674 (STRONGEST-Scalable, Tuneable and Resilient Optical Networks Guaranteeing Extremely-high Speed Transport).

8. References

Akar, N.; Eramo, V. & Raffaelli, C. (2011). Comparative analysis of power consumption in asynchronous wavelength modular optical switching fabrics. *Optical Switching and Networking*, Vol. 8, No. 3, July 2011, pp. 139-148, ISSN 1573-4277

Aleksic, S (2011). Energy Efficiency of Electronic and Optical Network Elements. *IEEE Journal of Selected Topics in Quantum Electronics*, Vol. 17, No. 2, March/April 2011, pp. 296-308, ISSN 1077-260X

Ben Yoo, S.J. (2011). Energy Efficiency in the Future Internet: The Role of Optical Packet Switching and Optical-Label Switching. *IEEE Journal of Selected Topics in Quantum Electronics*, Vol. 17, No. 2, March/April 2011, pp. 406-418, ISSN 1077-260X

Benes, V.E. (1965). *Mathematical Theory of Connecting Networks*, Academic Publishing, New York.

Chao, H.; Wu, U.; Zhang, Z.; Yang, S.; Wang, L.; Chai, Y.; Fan, J. & Choa, F. (2000). A photonic front-end processor in a WDM ATM multicast switch. *IEEE Journal of Lightwave Technology*, Vol. 8, No. 3, March 2000, pp. 273-285, ISSN 0733-8724

Ehrhardt, A.; Eiselt, M.; Groβkopf, G.; Kuller, L.; Ludwig, W.; Pieper, R.; Schnabel, R. & Weber, H. (1993). Semiconductor Laser Amplifier as Optical Switching Gate. *IEEE Journal of Lightwave Technology*, Vol. 11, No. 8, August 1993, pp. 1287-1295, ISSN 0733-8724

Eramo, V. (2000). Packet Loss in a Bufferless Optical WDM Switch Employing Shared Tunable Wavelength Converters. *IEEE Journal of Lightwave Technology*, Vol. 18, No. 12, December 2000, pp. 1818-1833, ISSN 0733-8724

Eramo, V.; Listanti, M.; Nuzman, C. & Whiting, P. (2002). Optical Switch Dimensioning and the Classical Occupancy Problem. *International Journal Communications Systems*, Vol. 15, No. 2, March/April 2002, pp. 127-141, ISSN 1074-5351

Eramo, V.; Listanti, M.; & Pacifici, P. (2003). A Comparison Study on the Number of Wavelength Converters Needed in Synchronous and Asynchronous All-Optical Switching Architectures. *IEEE Journal of Lightwave Technology*, Vol. 21, No. 2, February 2003, pp. 340-355, ISSN 0733-8724

Eramo, V. (2006). An Analytical Model for TOWC Dimensioning in a Multifiber Optical-Packet Switch. *IEEE Journal of Lightwave Technology*, Vol. 24, No. 12, December 2006, pp. 4799-4810, ISSN 0733-8724

Eramo, V.; Germoni, A.; Savi, M. & Raffaelli, C. (2008). Multifiber Shared-Per-Wavelength All-Optical Switching: Architectures, Control, and Performance. *IEEE Journal of Lightwave Technology*, Vol. 26, No. 5, March 2008, pp. 537-551, ISSN 0733-8724

Eramo, V.; Germoni, A.; Savi, M. & Raffaelli, C. (2009). Packet loss analysis of shared-per-wavelength multi-fiber all-optical switch with parallel scheduling. *Computer Networks*, Vol. 53, No. 2, February 2009, pp. 202-216, ISSN 1389-1286

Eramo, V.; Germoni, A.; Cianfrani, A.; Savi, M. & Raffaelli, C. (2009). Loss Analysis of Multiple Service Classes in Shared-per-Wavelength Optical Packet Switches. *IEEE/OSA Journal of Optical Communications and Networking* , Vol. 1, No. 2, July 2009, pp. A69-A80, ISSN 1943-0620

Eramo V. & Listanti M. (2009). Power Consumption in Bufferless Optical Packet Switches in SOA Technology. *IEEE/OSA Journal of Optical Communications and Networking* , Vol. 1, No. 3, August 2009, pp. B15-B29, ISSN 1943-0620

Eramo V. (2010). Comparison in Power Consumption of Synchronous and Asynchronous Optical Packet Switches. *IEEE Journal of Lightwave Technology*, Vol. 28, No. 5, March 2010, pp. 847-857, ISSN 0733-8724

Eramo, V.; Germoni, A.; Cianfrani, A.; Listanti, M. & Raffaelli, C. (2011). Evaluation of Power Consumption in Low Spatial Complexity Optical Switching Fabrics. *IEEE Journal of Selected Topics in Quantum Electronics*, Vol. 17, No. 2, March/April 2011, pp. 396-405, ISSN 1077-260X

Hinton, K.; Rakutti, G.; Farrel, P.; & Tucker, R.S. (2008). Switching Energy and Device Size Limits on Digital Photonic Signal Processing Technologies. *IEEE Journal of Selected Topics in Quantum Electronics*, Vol. 14, No. 3, May/June 2008, pp. 938-945, ISSN 1077-260X

Kalman, R.F.; Kazovsky, L.G. & Goodman J.W. (1992). Space Division Switches Based on Semiconductor Optical Amplifiers. *IEEE Photonic Technology Letters*, Vol. 4, No. 9, September 1992, pp. 1048-1051, ISSN 1045-1135

Mack, J.P.; Poulsen, H.N. & Blumenthal D.J. (2008). Variable Length Optical Packet Synchronizer. *IEEE Photonic Technology Letters*, Vol. 20, No. 14, July 2008, pp. 1252-1254, ISSN 1045-1135

Okada, A.; Sakamoto, T.; Sakai, Y.; Noguchi, K. & Matsuoka M. (2001). All-optical packet routing by an out-of-band optical label and wavelength conversion in a full-mesh network based on a cyclic-frequency AWG, *Proceedings of OFC 2001*, pp. ThG5, S. Diego (CA), March 2001

Sabella, R.; Listanti, M. & Eramo V. (2000). Architectural and technological issues for future optical Internet networks. *IEEE Communications Magazine*, Vol. 38, No. 9, September 2000, pp. 82-92, ISSN 0163-6804

Sakaguchi, J.; Salleras, F.; Nishimura K. & Ueno Y. (2007). Frequency-dependent Electric dc Power Consumption Model Including Quantum-Conversion Efficiencies in Ultrafast All-Optical Semiconductor Gates around 160 Gb/s. *Optics Express*, Vol. 15, No. 10, October 2007, pp. 14887-14900, ISSN 1094-4087

Sakamoto, T.; Okada, A.; Hirayama M.; Sakai Y.; Morikawi O.; Ogawa I.; Sato R.; Noguchi K. & Matsuoka M. (2002). Optical Packet Synchronizer using Wavelength and Space Switching. *IEEE Photonic Technology Letters*, Vol. 14, No. 9, June 2002, pp. 1360-1362, ISSN 1045-1135

Tucker, R.S. (2011). Green Optical Communications-Part II: Energy Limitations in Networks. *IEEE Journal of Selected Topics in Quantum Electronics*, Vol. 17, No. 2, March/April 2011, pp. 261-274, ISSN 1077-260X

Zucchelli, L.; Bella, D.; Fornuto, G.; Gambini, P.; Re, D.; Delorme, F.; Kraehenbuehl R. & Melchior, H. (1998). An experimental optical packet synchronizer with 100 ns range and 200 ps resolution, *Proceedings of ECOC 1998*, pp. 587-588, Madrid (Spain), September 1998

5

Multi-Functional SOAs in Microwave Photonic Systems

Eszter Udvary and Tibor Berceli
Budapest University of Technology and Economics,
Department of Broadband Infocommunications and Electromagnetic Theory,
Hungary

1. Introduction

The Semiconductor Optical Amplifier (SOA) is a very attractive device for optical communication systems because of their multi-functional capability. The operation of the SOA is controlled by both the electrical and optical input signal. The SOAs have demonstrated their multi-functional capability by combining optical amplification with modulation, gating, photo-detection, dispersion compensation, linearization, etc. The chapter describes the applications of SOA-modulator, SOA-detector and SOA-dispersion compensator in microwave photonic communication systems.

The design and construction of complex optical circuits exhibiting several functionalities are difficult tasks. Optical semiconductor integrated circuits having different functional elements on a single substrate have been developed and intensively studied. In that case individual functional elements need not be connected to each other through passive waveguides. Compared to a case when functional elements are independently formed, it is simpler to apply multifunctional devices. In this case a single device replaces numerous special elements. Multi-functionality in optical communication systems decreases complexity, reduces fabrication, installation and maintenance cost, minimizes the size, enhances the reliability and allows systems to work simultaneously with suitable parameters. However, we have to compromise, because the specialised devices have better operation parameters than multi-functional devices. Therefore, the degradation of the characteristics has to be minimized; hence the study of potential multi-functional devices is a very important task.

2. Radio-over-Fibre systems

Radio-over-Fibre (RoF) technology [Seeds] offers a perspective solution to the demand for wireless connection to the costumer („last or first mile problem"). It entails the use of optical fibre links to distribute RF signals from a central location to Remote Antenna Units (RAUs). It combines the properties of the microwave and photonics approaches. In narrowband communication systems and WLANs (wireless local area networks), RF signal processing functions such as frequency up-conversion, carrier modulation, and multiplexing, are performed at the radio base station. RoF makes possible to centralise the RF signal processing functions in one shared location, and then to use optical fibre, which offers low signal loss (0.3 dB/km for 1550 nm, and 0.5 dB/km for 1310 nm wavelengths) to distribute the RF signals to the RAUs. This way the RAUs are simplified significantly, as they only need to perform optoelectronic conversion and amplification functions. For broadband services the frequencies are in the millimetre wave range (like 60 GHz fibre radio link). Such a concept is based upon an optical link between the central station and the RAU in a picocellular structure [Ng'oma]. Some of the advantages and benefits of the RoF technology compared with electronic signal distribution are the following: low attenuation loss, large bandwidth, immunity to radio frequency interference, easy installation and maintenance, multi-operator and multi-service operation, and dynamic resource allocation. These benefits can translate into major system installation and operational savings, especially in wide-coverage broadband wireless communication systems, where a high density of RAUs is necessary.

The SOA is a potential candidate for an electro-optic transceiver (transmitter and receiver) in a RoF network. The SOA operates as a modulator to add a new channel, as a detector to drop the needed channel and as an in-line amplifier to amplify the other channels, simultaneously. It realizes a compact, small size and cost-effective radio repeater for signal distribution. Fig.1. shows a simplified system setup representing the SOA based RAU. The output power of the laser source is modulated with the 1st information channel applying a SOA-modulator. The SOA transceiver detects the 1st channel and adds the 2nd channel in the 1st stage. In the 2nd stage the SOA transceiver adds the 3rd channel, but it can detect the 1st or 2nd channel selecting by tuneable filter. Finally the receiver side every information channels can be extracted. The information channels and the bias of the SOA are separated by bias tee circuit. The separation of add and drop channels can be achieved in the electrical regime by an electrical branching filter (1st stage) or an electrical circulator with bandpass filter (2nd stage). In the first case the realisation of a reconfigurable add/drop multiplexer is difficult, in the second case an electrical filter is needed also. In the drop branch a high power RF amplifier provides the suitable power at the antenna output, and a low noise amplifier is necessary in the add branch.

Other advanced devices like electro-absorption transceiver for signal remodulation or polarization rotation remodulator have been designed though only modulation is executed, whilst with semiconductor amplifiers it's possible to perform amplification, modulation and detection with the same optical device. The signal distribution of RoF systems can be realised by point-to-point, point-to-multipoint, bus, ring and open loop topologies.

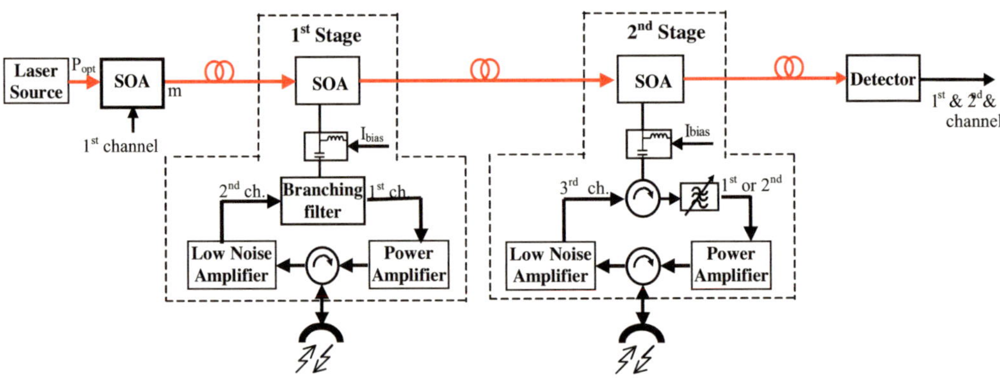

Fig. 1. Simplified system setup with SOA transceivers

2.1 Point-to-point and star topology

The Point-to-point topology is the simplest method; it is a permanent link between two endpoints. In a star topology, each RAU is connected to a central unit with a point-to-point connection. The star topology is considered the easiest topology to design and implement. Adding additional nodes is simple, but the central unit represents a single point of failure. Fig. 2. represents the star topology of RoF system. Bi-directional data transmission can be achieved over duplex optical fibre. On the other hand the uplink and downlink can be separated by optical circulator. Anyhow, the RAU consist optical source for electrical-to-optical conversion.

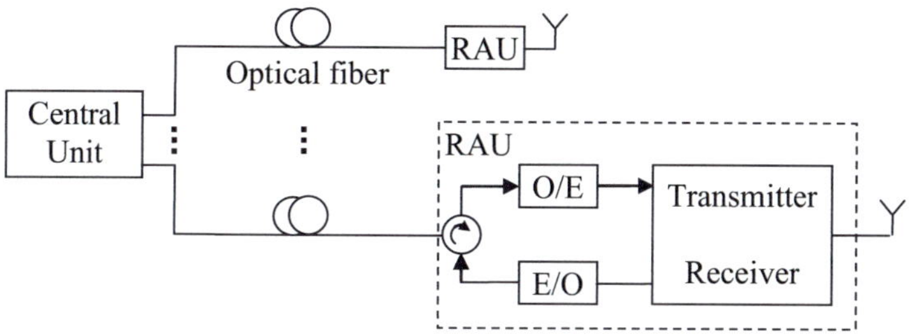

Fig. 2. Star topology, RoF system with traditional Electrical-to-Optical (E/O) and Optical-to-Electrical (O/E) converters

For more simplified RAU structure, the laser source of uplink optical carrier can be moved to the central unit and the continuous wave (CW) optical signal is transmitted over the optical fibre to the RAU. So, there are no injection of light is added at the RAU for the uplink transmission. The optical carrier is attenuated; hence some amplification should be required, in order to arrive to the maximum number of users and larger distances. Bidirectional amplification is possible by EDFAs (erbium doped fiber amplifiers) but the price is high and an additional high pump power is required. In this way, SOAs are suitable to be positioned at the simplified RAU. They not only carry out the modulation and detection but also offer an additional optical gain.

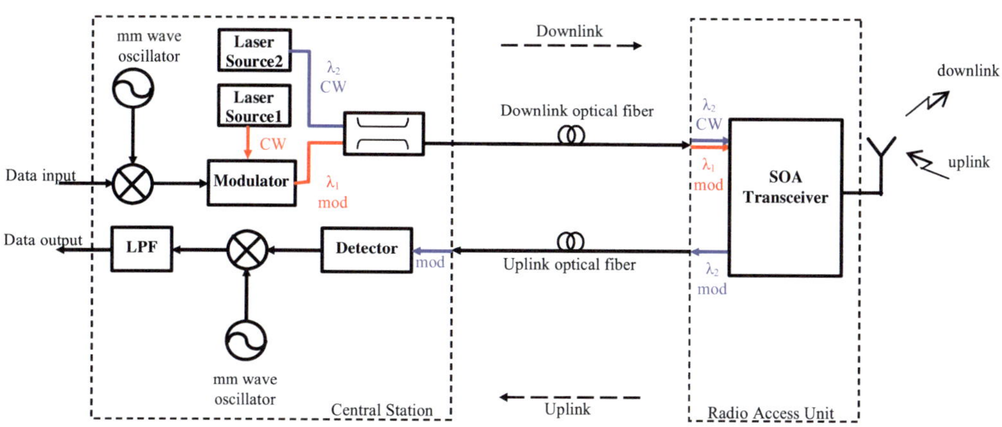

Fig. 3. Point-to-point topology RoF system applying SOA transceiver and centralised laser sources

2.2 Ring and open loop topologies

The ring topology means, that all stages are connected to one another in the shape of a closed loop, so that each stage is connected directly to two others, one on either side of it. The ring or loop network topologies are set up in a circular fashion in which data travels around the ring in one direction and each unit on the right acts as a repeater to keep the signal strong as it travels. Each stage incorporates a receiver for the incoming signal and a transmitter to send the data on to the next device in the ring. The open loop topology is not closed; it is similar to bus topology, where each node is connected to a single fibre.

Fig.4. shows a loop topology for signal distribution in a RoF system. In the network the transmission frequencies are fixed and the reception frequencies are different at each node. A control unit provides information for the nodes concerning the subcarrier frequency to be received. In the ring type version of the network a single fiber is used at each node both for transmission and reception of the information. The ring structure enables each node to communicate with any other one. In the open loop version of a network the optical fiber collects the information of the nodes and then it is routed back to make possible the reception of the collected information. Applying that folded back construction each node can communicate with any other one. However, the distance between the nodes is limited by the fiber loss and the available power.

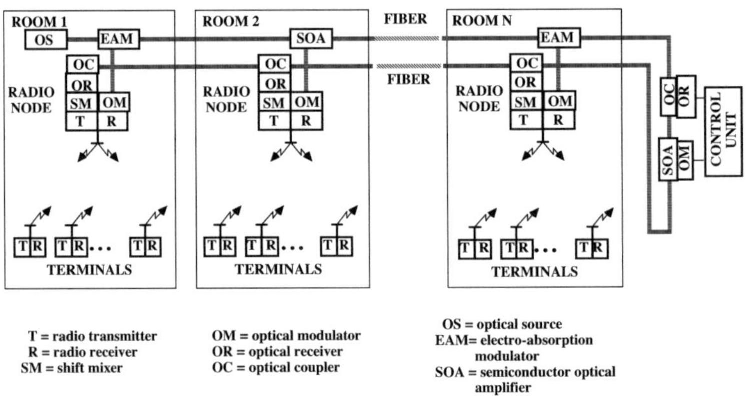

Fig. 4. Loop topology with traditional Electrical-to-Optical (E/O) and Optical-to-Electrical (O/E) converters

In Fig.5. the open loop version of a WDM-RoF network is shown. In the combined WDM-RoF network every node has a fix optical carrier instead of electrical subcarrier. In this way, the transmission capacity is extended. The multifunctional SOA can operate as a transparent unidirectional add-drop node with potential applications in photonic ring, loop or open loop networks, whose main feature is the possibility of adding new nodes and/or increasing the bit-rate for the required nodes without reconfiguring the whole network. The concept of the complete system is based on the cascaded SOA chain approach which produces simultaneous modulation and amplification. Furthermore, such read-write nodes need a minimum amount of external electronics, therefore reducing network cost and complexity. The Optical Add Drop Multiplexers (OADMs) select the dedicated optical carriers for the transceivers. The multifunctional SOA detects the information from the downlink optical carrier and modulates the uplink optical carrier with the information received from the antenna.

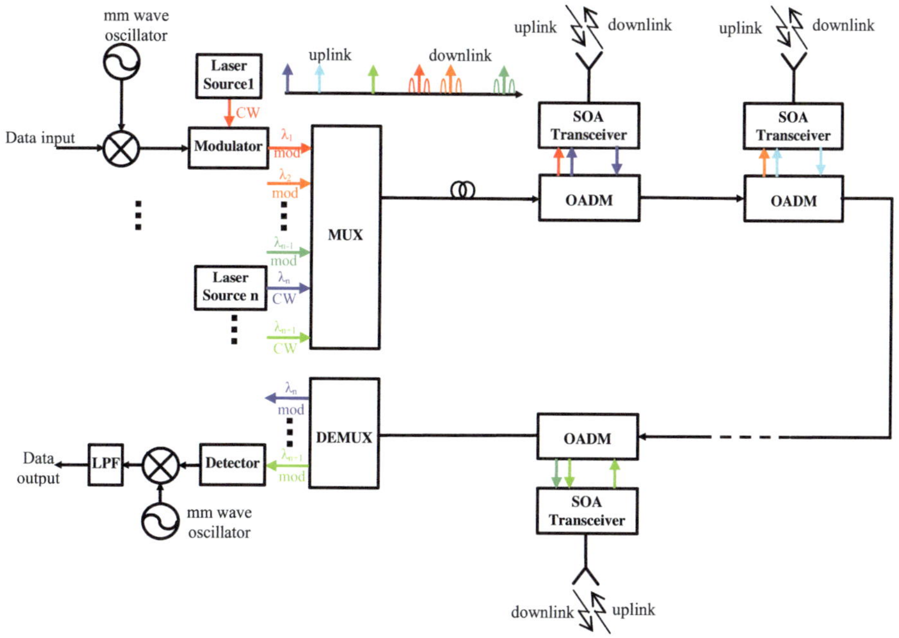

Fig. 5. WDM - RoF open loop concept with SOA transceivers

2.3 Hybrid WDM/SCM passive optical network

Fiber-to-the-home (FTTH) technology is one of the main research objectives in the "broad-band for all" concept that encourages the development of optical access infrastructure. In order to fulfil this concept, cost effective solutions must be developed to be able to offer future-proof broad-band connections to end users at a reasonable cost [Arellano]. A key element in access networks is the optical network unit (ONU) of the customer premises equipment, having a direct impact on the cost per customer, whereas, the access part represents the main segment of the total capital cost; thus, simple ONUs need to be designed. Other key desirable characteristics of an access network are the use of one single fiber for both upstream and downstream transmission in order to reduce network size and connection complexity of the outside plant [Prat], the elimination of the laser source at the ONU, thus avoiding its stabilization and provisioning, and, if possible, all ONUs being wavelength independent, to fit in a transparent wavelength-division-multiplexing scenario of a future FTTH (fiber-to-the-home) network. The SOA-modulator-detector may be used in a basic bidirectional single-fiber single-wavelength scenario for a FTTH network. However it works as a point-to-point connection, hence a reflective structure (RSOA) is more powerful.

The present passive optical network (PON) is standardised. The future, upgraded systems are waiting for standardisation. Wavelength division multiplexed-passive optical network (WDM-PON) is a promising solution for the future high-speed access networks such as FTTH or fiber to the office (FTTO) by reasons of large capacity, network security, protocol transparency and upgradability [Kang]. However, because of relatively expensive WDM components, the WDM-PON has been considered as a next-generation. Recently, to overcome this problem, there have been several proposals. Normally, two wavelength

sources are required for both up- and down-link transmission. But, one approach is re-modulation of down stream signal at ONU for upstream transmission.
The downstream signal modulates directly a laser diode, in upstream transmission the downstream signal re-modulates the optical carrier using a SOA with the SubCarrier Multiplexed (SCM) technique. No additional high cost devices are required such as external modulator and optical amplifier. A WDM-PON employing a SOA as a modulator has some advantages. The SOA gives additional gain for incident optical power to overcome device and transmission losses. Hence the SOA may be used as a modulator which accomplishes both modulation and amplification (Fig.6.). The SOA which operated in gain saturation region can reduce the intensity noise of optical signal. Due to the mixture of WDM and SCM techniques, a simple ONU which shares the same wavelength both up- and down-link transmission is possible [Kang].

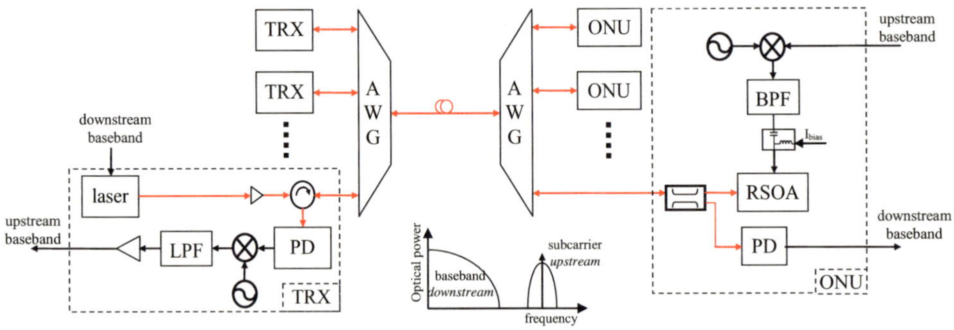

Fig. 6. Hybrid WDM/SCM PON with multifunctional SOA

3. Multi-functionality

The investigation of potential multifunctional devices is very important task for the realisation study of the proposed system concepts. Multifunctional facilities of SOAs are presented by combining optical amplification with modulation and photo-detection.

3.1 Amplification

There are several applications of SOAs in today and the future analogue and digital optical networks. The optical losses are compensated by optical amplifiers (OAs), it is well used in loss limited systems as post, in-line or pre-amplifier. The SOA is a semiconductor based, small size, compact, low cost, current driven device, which amplifies the incoming optical signal directly in the optical regime, without any optical/electrical conversion. Moreover, the semiconductor technology offers a wide flexibility in the choice of the gain peak wavelength by just appropriately choosing the material composition of the active layer. Another key advantage is that these devices can be integrated with other active or passive optical components to generate more complex functionalities. Finally, they are potentially cheap, thanks to the mature technology basis.

3.2 Detection function

The detection means optical to electrical conversion. The optical detector is characterized by the responsivity, the sensitivity and frequency response. The SOA-detector responsivity can be determined from the variation in the injection current. The magnitude and purity of the detected signal depend on the modulation signal, the bias current, the input power and the operation parameters of the SOA [Udvary1].

3.2.1 Operation principles

Two different mechanisms induce detection in the SOA. Operated at an injection current corresponding to an electron density below transparency, the device works as a photo-detector and the detection signal arises due to absorption of the injected light and the creation of electron-hole pairs. However at injection current above transparency, that is the amplifying regime, the injected optical signal will cause stimulated transitions, which will reduce the carrier density in the gain medium. Due to these two different mechanisms of interaction, the detected electrical signal will change polarity at transparency (Fig.7.).

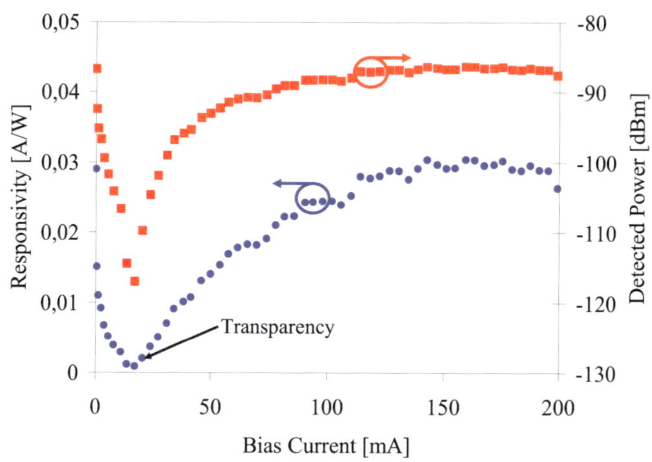

Fig. 7. Detection behaviour of SOA. Measured detected power and detection responsivity versus operation point of the SOA, optical wavelength=1550nm, Optical power at the input of the SOA-detector=40μW, Modulation depth=20%, Temperature=20°C

3.2.2 Static characteristics

The static photo-detection current is not given data in data sheets of traditionally used SOAs. In unsaturated regime this curve can be approximate calculated from the optical gain (proportional with G-1) and the input optical power (proportional with Pin). However in several cases it can only be determined by measurements.

Fig.8a depicts the electrical current as a function of the input optical power of SOA-detector. The relationship between the input optical power and the output electrical current is given by the detector responsivity, it can be determined from this curve. However this static photo-detection current is operation point and temperature sensitive. Fig.8b represents the shape of the calculated curve (based on optical gain-bias current curve), that follows the measured data.

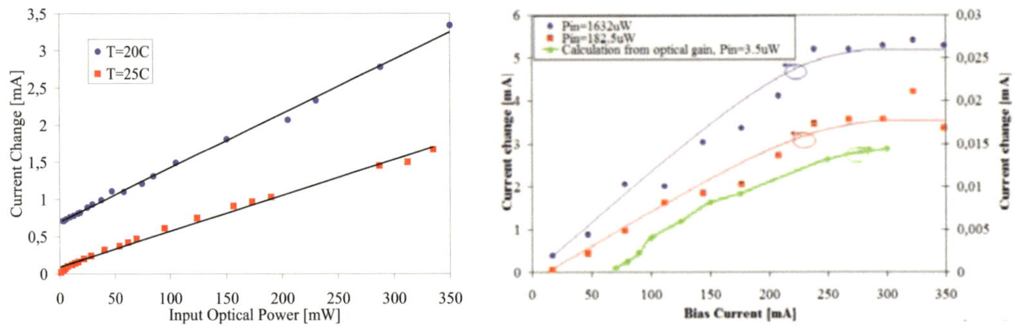

Fig. 8. Static detection characteristics of SOA-detector
Optical wavelength=1550nm, Applying isolators to eliminate the optical reflection effect
(a) Measured current change versus average optical power, at different temperatures, Bias Current=150mA
(b) Current change versus SOA operation point for different input optical power, Comparison of measurement and calculation results, Temperature=20°C

3.2.3 Detection of intensity modulated signal

The detection functionality is studied using the following link (Fig.9.). The optical signal is intensity modulated. The simplified small signal calculation applies sinusoidal modulation part. The optical link attenuates the optical signal and the SOA-detector receives it. The modulation information appears at the electrical connection of the SOA-detector.

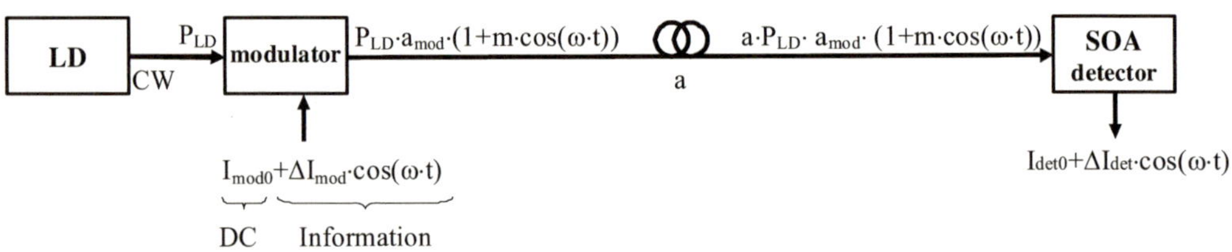

Fig. 9. Simplified block diagram for calculation of SOA-detector parameters

The detection functionality was validated by experimental work. During the measurements the SOA-detector under test was driven by the bias (dc) current and a bias tee separated the detected sinusoidal modulation signal. The polarization state of the incoming optical power was optimized by a polarization controller because the measured SOAs were polarization sensitive. The harmful effect of the optical reflection was eliminated by optical isolators. The required optical power and wavelength were produced by a tunable laser source. The intensity modulated optical signal was generated by a Mach-Zehnder modulator (MZM). The setup was controlled by a computer program, hence the measurement parameters were carefully set by the program and the measurement results were processed and restored. The detected electrical power was measured with different parameters.
When the bias current increases, i.e. the population inversion and the gain are higher the detected power increases (Fig.10). The diagram follows the shape of the optical gain curve. So the optimum bias current is the same from the viewpoint of the detection and amplification functions. The two curves have reference to two different modulation depths

of the incoming intensity modulated signal. The detected current is directly proportional to the modulation depth. Thus, the difference between the two electrical powers is about 14dB.

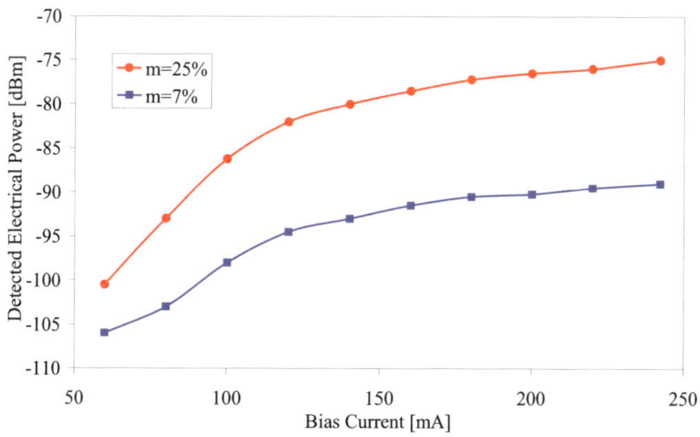

Fig. 10. Measured detected electrical power versus operation point of SOA-detector with different modulation depths, Optical power at the detector input=5μW, Modulation frequency=3GHz, Optical wavelength=1550nm, Temperature=20°C, Applying isolators to eliminate the optical reflection effect

The result of the detection experiments over input optical power and temperature are depicted in Fig.11. In the unsaturated regime the detected electrical signal is square proportional to the optical power, but in the saturation regime the relation goes to linear proportionality. The detection is also temperature sensitive, because the operation of semiconductor devices depends on the temperature. The measurement results show, that the detection efficiency decreases, when the temperature increases.

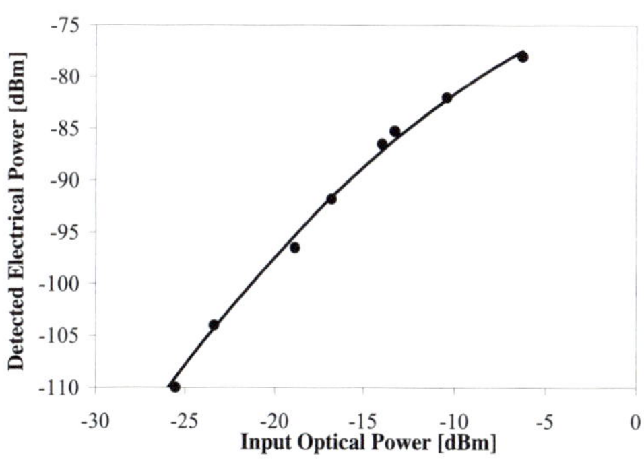

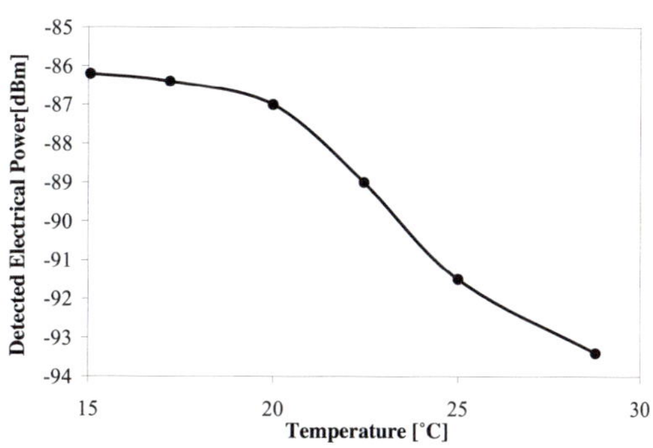

Fig. 11. Measured detected electrical power versus average input optical power (a) and temperature (b). Optical wavelength=1550nm, Bias current=200mA, Modulation frequency=3GHz, Modulation depth=25%, Isolators

3.2.4 Responsivity extraction from the measurements

The responsivity of the SOA-detector can be calculated from the measurement. If the input signal is intensity modulated, the fluctuation in the optical intensity due to modulation will induce fluctuation in the injection current. We shall consider sinusoidal intensity modulated input optical signal

$$P_{in}(t) = a \cdot a_{mod} \cdot P_{LD} \cdot (1 + m \cdot \cos \omega t) = P_{DC} + \Delta P^{opt}_{SOA\_in} \tag{1}$$

where P_{in} is the input optical power of SOA-detector, P_{LD} is the average optical power of the laser, a_{mod} is the optical loss of the modulator, a is the optical loss between the modulator and SOA-detector, m is the modulation depth and ω is the angular modulation frequency. Hence the detected electrical current has cosine type component

$$I_{det\,SOA} = I_{DC} + \Delta I_{det\,SOA} \cdot \cos(\omega t) \tag{2}$$

Where I_{detSOA} is the current detected by SOA, I_{DC} is the average detected current, ΔI_{detSOA} is amplitude of the detected cosinusoidal signal.

The SOA-detector responsivity (R) can be computed from the detected electrical power (P_{detSOA})

$$\Delta I_{det\,SOA} = R \cdot m \cdot \frac{P_{DC}}{a_{in}} \quad \Rightarrow \quad R = \sqrt{\frac{2 \cdot a_{in}^2 \cdot P_{det\,SOA}}{m^2 \cdot P_{DC}^2 \cdot Z}} \tag{3}$$

where Z is the microwave impedance of SOA.

3.2.5 Noise calculation

The optical output from an optical amplifier is composed of an amplified optical signal and an amplified spontaneous emission (ASE) with broad spectral width. Moreover interference is created between ASE components and light signal [Shiraz]. So several types of noises (the shot noise belonging to signal and spontaneous emissions, beat noise between signal and spontaneous emissions, beat noise between spontaneous emission components, thermal noise of the receiver and excess noise belonging to incoherence of the input signal) can be observed, when the output photons are detected by a photo-detector. In the unsaturated region the beat noise between the ASE components dominates. The installation of an optical bandpass filter into the microwave photonics link decreases the noise bandwidth and the equivalent noise power. In SCM system only a single optical carrier is applied, hence small optical bandwidth (1-2nm) should be specified in the link for decreasing the noise build-up.

The noise generated by a SOA acting as a detector is different from the noise generated by a SOA acting as an amplifier or modulator. There are several similar contributions to the total noise power at electrical connection of SOA-detector. However interference between ASE components and light signal is created inside the SOA device. So the noise components originating from the light amplification depend on input parameters as wavelength, input power or driving current with different aspect, than in case of other applications. Shot noise caused by random generation and flow of mobile charge carriers and thermal noise of the load resistor are well known and can be expressed like in a traditional PIN detector [Agraval]. The thermal noise due to electrical amplification has been included as well [Gustavsson].

$$i_{shot}^2 = 2 \cdot e \cdot I \cdot B_0, \qquad i_{therm}^2 = \left[\frac{1}{R_L} + \frac{1-F}{R_N}\right] \cdot 4 \cdot k \cdot T \cdot B_0 \tag{4}$$

where R_N is a standard 50Ω resistance, F is the electrical amplifier noise figure, k is the Boltzmann constant, T is the temperature, B_0 is the detection bandwidth.
The mean value and the variance of the number of output photons per second due to light amplification process can be calculated. The variance of the detected current due to the photon noise can be obtained from the variance of photons in the amplifier medium. The generated current is proportional to the photon density in the cavity [Gustavsson].

$$i_{ph}^2 = e^2 \cdot \left(\Gamma \cdot g_m \cdot L\right)^2 \cdot \overline{\sigma}^2 \cdot B_0 \tag{5}$$

where L is the device length, $\overline{\sigma}^2$ is the photon variance averaged over the amplifier length. In the case of a true traveling wave amplifier (the face reflections are zero) the equivalent noise bandwidth for the beat noise between spontaneous emission components and the equivalent noise bandwidth for spontaneous emission shot noise are equal (Δf) and the variance of the detected current can be calculated [Gustavsson].

$$i_{ph}^2 = e^2 \cdot \left(\Gamma \cdot g_m \cdot L\right)^2 \cdot \frac{(G-1)^2}{\ln G} \cdot B_0 \cdot \begin{bmatrix} 2 \cdot n_{sp} \cdot \dfrac{\lambda}{h \cdot c} \cdot P_{in} + \dfrac{2}{G-1} \cdot \dfrac{\lambda}{h \cdot c} \cdot P_{in} + \\ +n_{sp}^2 \cdot \Delta f \cdot \left(1 - 2 \cdot \dfrac{G - \ln G - 1}{(G-1)^2}\right) + \\ +2 \cdot n_{sp} \cdot \Delta f \cdot \dfrac{G - \ln G - 1}{(G-1)^2} \end{bmatrix} \tag{6}$$

where n_{sp} is the population inversion parameter, $\langle n_{in} \rangle$ is the mean value of the number of input photons per second. The first term of equation represents the beat noise between the signal and the spontaneous mode, the second term represents signal shot noise, the third term represents spontaneous- spontaneous beat noise, and the last term represents spontaneous shot noise over the entire amplifier spectrum.
The signal-to-noise ratio is given by the following equation:

$$SNR = \frac{i_{sig}^2}{i_{ph}^2 + i_{therm}^2} \tag{7}$$

It is of interest to determine the magnitude of each contribution to the total noise in order to see which component dominates at different system parameters. The different noise components depend on the optical signal level with different aspect. The thermal noise, the spontaneous shot noise and the spontaneous beat noise are independent in the unsaturated regime, the signal shot noise and signal- spontaneous beat noise have linear relation with the input optical power. In case of small input optical power one of the constant noises dominates. Then signal shot noise or signal- spontaneous beat noise overcomes this limit. These relationships are illustrated in Fig.12 in which the different noise components and the total noise are calculated as a function of the link loss or input optical power of the SOA-detector. The calculation uses the measured SOA parameters and takes into account the gain saturation effect. Hence the spontaneous shot noise and the spontaneous beat noise start to decrease as the optical gain decreases.
Similar results can be observed in case of constant input optical power as a function of the SOA gain (Fig.12). For low gain values thermal noise and shot noise dominate and for larger

gain the beat noises give the dominating contribution. So the signal-to-noise ratio (SNR) increases up to a certain gain, reaches a maximum, finally it decreases. There exists an optimum amplifier bias point corresponding to maximum SNR of the SOA-detector. Although higher gains would yield higher responsivity, the SNR decreases and it is not desirable for the system.

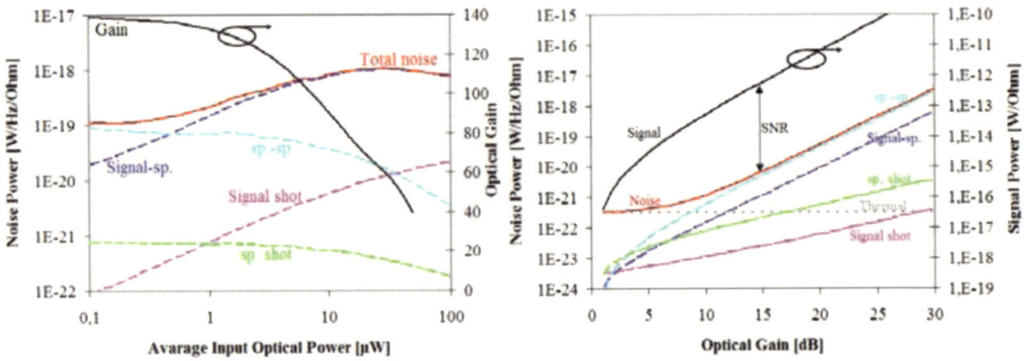

Fig. 12. Calculated noise and signal powers versus input optical power (a) and optical gain (b), calculation based on measured data

3.3 Modulation function

The optical gain of the amplifier depends on the bias current and taking advantage of this effect could be used as an external modulator. The bias current of the SOA is modulated; therefore the material gain and the intensity of the output signal are modulated [Mork]. If small signal current modulation is considered, the electrical signal contains an invariant and a modulation parts, hence the number of charge carriers and photons are also time dependent [Conelly]. The magnitude and purity of the signal depend on the modulation signal, the bias current, the input power and the operation parameters. The SOA modulator requires low modulation power, and the detected electrical power is high because of the optical gain in contrary to the optical insertion loss of other modulators. However the SOA has noticeable optical noise [Udvary1].

3.3.1 Calculation of the realizable modulation depth

The modulation functionality is studied with the following link (Fig.13.). The optical signal from the laser diode is intensity modulated by SOA with time dependent optical gain. The intensity modulated optical signal is detected by traditional PIN photodiode.

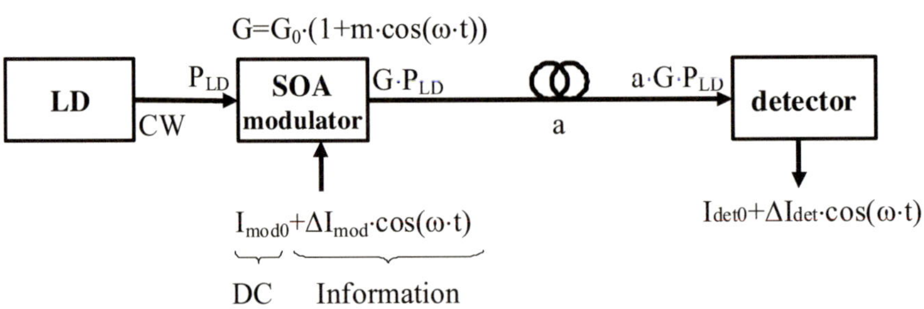

Fig. 13. Simplified block diagram of the link for calculation of modulation behaviour

The modulation operation can be derived based on the slope of the measured optical gain curve (m_d) and the average optical gain (G_0).
The current and the gain of the device are:

$$I(t) = I_0 + \Delta I_{mod} \cdot cos(\omega t) \qquad G(t) = G_0 + \Delta G \cdot cos(\omega t) \tag{8}$$

where I_0 is the constant (dc) current, ΔI is the current modulation amplitude, G_0 is the constant optical gain of SOA, ΔG is the modulation part. Hence the optical signal at the output of the SOA-modulator takes the form

$$P_{out} = G_0 \cdot P_{in} \cdot (1 + m \cdot cos(\omega t)) \tag{9}$$

where the modulation index (m) is

$$m = \frac{\Delta G}{G_0} = \frac{m_d \cdot \Delta I_{mod}}{G_0} = \frac{m_d}{G_0} \cdot \sqrt{\frac{2 \cdot P_{mod}}{Z}} \tag{10}$$

where P_{mod} is the modulation electrical power, Z is the microwave impedance of SOA.
The output signal of the SOA-modulator is detected by an optical-electrical converter, P_{det} is the detected electrical modulation power

$$P_{det} = \eta^2 \cdot \frac{P_{in}^2}{a^2} \cdot m_d^2 \cdot P_{mod} \tag{11}$$

η is the detection efficiency, a is the optical loss between the SOA and the detector.
The modulation depth is proportional to the slope of the gain curve and the electrical modulation power, but it is in inverse relation to the average optical gain (Eq.16). However, the detected electrical power increases with the modulation power (direct relation), the slope of the gain curve and the input optical power of the SOA-modulator (quadratic relation) increase (Eq.17). Same conclusions can be observed from the experiments.
Naturally, the modulation depth realized by the SOA-modulator can be computed from the measured detected electrical power with the knowledge of the modulation power and the input average optical power of the detector:

$$m = \frac{2 \cdot a^2 \cdot P\,det}{P_{in}^2 \cdot G_0^2 \cdot \eta_{det}^2 \cdot Z} = \frac{2 \cdot P_{det}}{P_{det}^{opt} \cdot \eta_{det}^2 \cdot Z} \tag{12}$$

Where P_{det}^{opt} is the average optical power at the input of the detector.

Based on this calculation the experimentally realized optical modulation depth is 5-10 percent with sufficiently low (-30dBm) modulation power.

3.3.2 Intensity modulation

Practically, the average optical gain and the slope of the optical gain - bias current curve determine the optimum working state of the SOA as a modulator. The curve can be divided into three parts (Fig.14.). In the first one the amplification just starts and it is not effective, the second one is the almost linear region and after it the slope of the optical gain starts to decrease. The middle of linear region of this curve should be chosen for operation point, because of the low static non-linear distortion effect and the high slope [Udvary2].
Fig.14. shows the measured optical gain and modulation behavior of the SOA-modulator versus bias current. The three regions are well seen in this figure, too. The injection current

is not enough for the expected work and the detected power is low in the first part. The power is near constant in the second linear part and after it the detected product starts to decrease because the slope of the gain curve falls. The lower curve at modulation behavior figure represents the result without input optical power, i.e. just the amplified spontaneous emission power (ASE) produces the modulated signal and the broadband O/E (optical-to-electronic) converter can detect this poor fluctuation. This effect can be dramatically decreased by a narrow band optical bandpass filter.

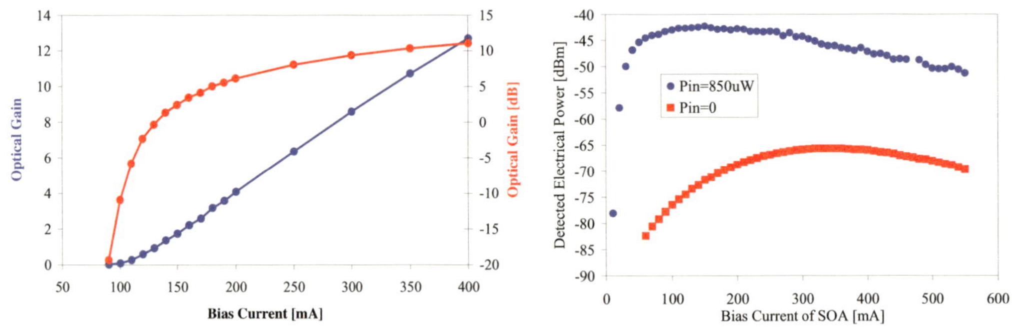

Fig. 14. Measured Optical Gain (a) and Modulation behaviour (b) of SOA-modulator versus bias current of the device. Wavelength=1550nm, Temperature=20°C, Modulation power=-30dBm, Modulation frequency=400MHz

The modulation bandwidth is limited by the speed at witch the carrier density can be changed, this is usually limited by the spontaneous lifetime of the carriers in the SOA (in the nanosecond range). The lifetime in the presence of a strong, saturating input signal is reduced due to stimulated recombination. The real speed depends on the structure of the device, but in general it is larger than 10GHz speed.

3.3.3 Linearity

Degradation of the transmission system will occur due to the crosstalk between the subcarriers (nonlinearity) and noise expansion (ASE). In linear regime the SOA modulator shows low nonlinearity because the noise generated by the SOA will dominate in the system. The intermodulation products overcome the noise floor in case of extraordinary high modulation indices and in the saturated operating region.

The nonlinearity causes intermodulation by different order mixing products. In practice, it is assumed that the transfer function of the device has only linear and cubic terms, because the even order terms do not produce mixing products falling into the transmission band and from order 5 the level of mixing products supposed to be very small.

In the two-tone intermodulation experiments the SOA was biased and modulated by the sum of two microwave signals. The output noise (P_{noise}) and signal levels were measured for the fundamental (P_1), the second (P_{sec}) and the third (P_{th}) order mixing products.

For characterizing the nonlinearity the third order intercept point (IP3) or the spurious suppression in dBc is used. However, for high quality signal transmission a high linearity is not sufficient because the noise has to be low as well. Therefore, the spurious free dynamic range (SFDR) is a better characteristic. It is dependent both on the linearity and noise, it is higher when the linearity is high and the noise is small. In personal communication systems

$72-83\ \mathrm{dB}\cdot\mathrm{Hz}^{2/3}$ SFDR is required [Olshansky]. The determination of SFDR, IP2 and IP3 are presented in Fig.15.

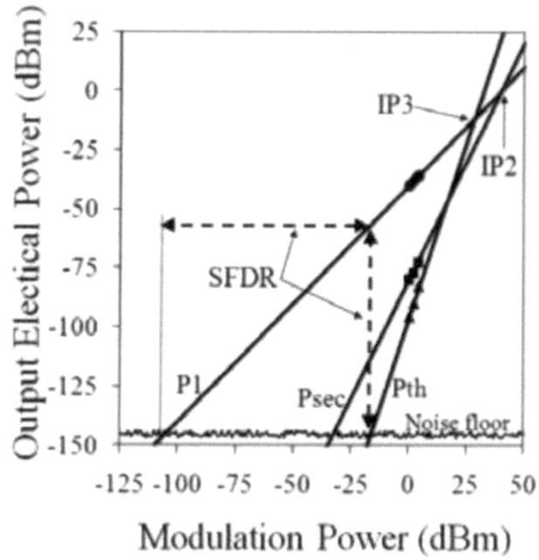

Fig. 15. Determination of SFDR, IP2, IP3.

Modulation frequencies=199, 200MHz, Modulation Power=4dBm, Input Optical Power=1mW, without Isolators, Temperature=20°C

$$IP2[dBm] = 2\cdot P_1[dBm] - P_{sec}[dBm]$$

$$IP3[dBm] = \frac{1}{2}\cdot\left(3\cdot P_1[dBm] - P_{th}[dBm]\right)$$

$$SFDR = \frac{P_{in}(P_{th} = P_{noise})}{P_{in}(P_1 = P_{noise})} = \frac{P_1(P_{th} = P_{noise})}{P_{noise}}$$

$$SFDR[dB] = \frac{2}{3}\cdot\left(IP3[dBm] - P_{noise}[dBm]\right)$$

The experimental work was done on different types of SOAs. The presented results characterize a commercial SOA having 13 dB of small signal gain, 15dBm of saturation power, 100nm of optical bandwidth, 300mA of bias current. All the measuring instruments were checked to have higher dynamic range and better linearity than the value expected from the SOA-modulator. 7% modulation depth was applied, because the modulation indices are usually less than 10 % in typical SCM systems. However, it was also checked for a wide range of modulation depth (3-30%).

The nonlinear behavior depends on several parameters [Marozsak]. Fig.16. shows the noise level, IP3 and SFDR versus bias current. The graph can be divided into three parts. First, the SOA is near the transparency, the nonlinearity is high; hence the IP3 and the SFDR improve as it approached the near linear range. In the second part the modulation and nonlinear products don't change significantly but the noise level rises, hence the SFDR decreases. Finally, the intermodulation products start rising and the degradation of the SFDR is faster. The shapes of the curves are similar for the results estimated from the one tone simulations. The reason of some difference (some dB) in the exact values is that for the measurements the SOA device's internal parameters were not available. So we could only use estimated parameters for the simulations.

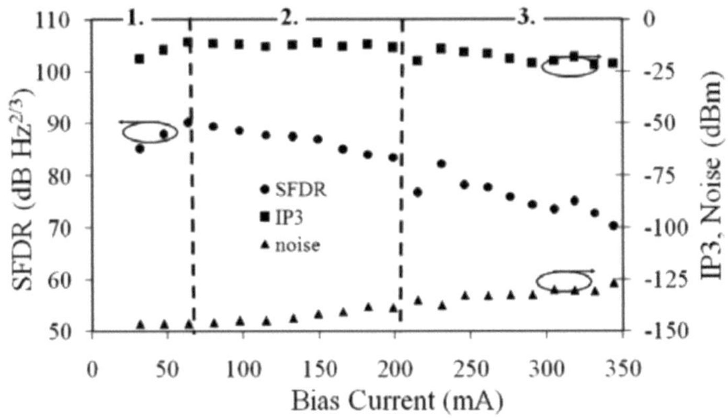

Fig. 16. Nonlinear behavior of SOA modulator versus bias point. Modulation frequencies=199, 200MHz, Modulation Power=4dBm, Input Optical Power=1mW, without Isolators, Temperature=20°C

By injecting more optical power, the noise floor reduces due to the amplified spontaneous emission (ASE) reduction in saturated regime. Same time the optical gain – bias current curve is more linear. The combination of these two factors allows us to obtain a better SFDR at high input optical power.

The linearity is temperature sensitive, because the operation of semiconductor devices depends on the temperature. The degradations of SFDR and IP3 are about $2\text{dB}\cdot\text{Hz}^{2/3}$ and 3dB for 10°C temperature change (Fig.17.).

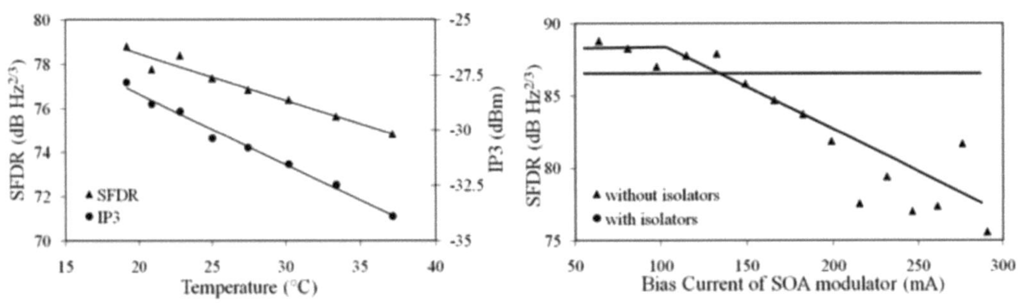

Fig. 17. Nonlinearity dependence on the temperature and optical reflection Wavelength=1550nm, Input Optical Power=1mW, Bias Current=50mA, Modulation frequencies=199, 200MHz

In short distance RoF link, the level of optical reflection is usually determined by the optical detector. The system will be more instable in case of strong optical reflection (without optical isolators), and larger SFDR degradation can be observed as seen in the Fig.17. The change of the SFDR is caused by two different effects (Fig.18.). The noise level of the device increases as a function of the bias point, the degradation is more significant without optical isolator. On the other hand the level of the nonlinear product will fluctuate in case of strong reflection.

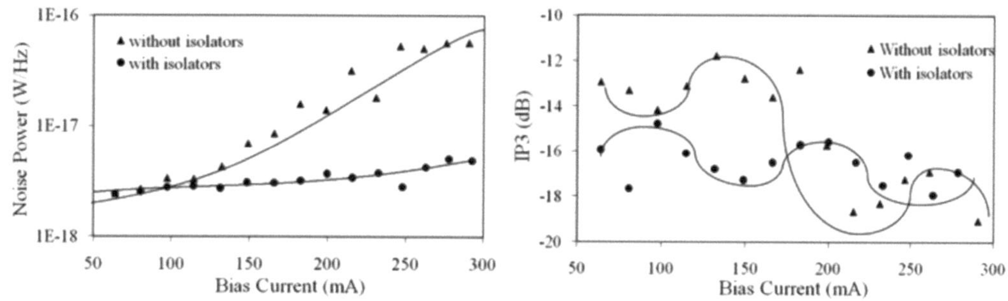

Fig. 18. Levels of noise and intermodulation products depends on the optical reflection. Wavelength=1550nm, Frequencies=199, 200MHz, Input Optical Power=1mW, Temperature=20°C

3.4 Dispersion compensation in RoF systems

Compensating dispersion penalty is a key problem when next generation Radio-over-Fiber networks are built. Several techniques have been proposed to overcome dispersion effect. An alternative method is presented to overcome the RF carrier suppression effect in optical links based on the joint effect of SOA chirp, chromatic dispersion and nonlinearities of the optical fiber. The results show that the frequency notches caused by the dispersion-induced carrier suppression effect may be sharply alleviated and the performance of the transmitted digital signal can be improved.
The SOA dispersion compensator has the advantage that it is a loss-less, wide band solution with robust operation. It is more efficient than midway optical phase conjugation or self phase modulation effect introduced by the fiber. It offers optical amplification compared with high insertion loss of dispersion compensation fiber. It has high bandwidth (30-35nm), hence it is transparent for optical or electrical carrier variation and more insensitive for environmental and system parameters than Fiber Bragg Grating. It is semiconductor based device, which can be easily integrated with semiconductor optical source. So it doesn't demand expensive and complex optical device (like SSB Mach-Zehnder modulator), just an additional integrated section is necessary in the optical source. Additionally the operation of the device can easily be optimized by bias point and input optical power control.

3.4.1 Basic operation

When the incoming optical power of the laser amplifier is intensity modulated, the optical gain is affected in both magnitude and phase via the modulation of the complex refractive index caused by the electron density. Consequently, in SOA the optical signal becomes amplitude modulated (AM) and phase modulated (PM). It can be modelled using the Linewidth Enhancement Factor (LEF=Henry factor=α factor) approximation. Measurements of LEF can are found in the literature and have shown that LEF is not a mere constant factor, but it is for instance a function of bias current, wavelength and input optical power. In the unsaturated region the LEF value ranges from 2 to 7 for GaAs and GaInAsP conventional lasers and from 1.5 to 2 for quantum well lasers [Occhi]. However, the chirping parameter which is positive for light sources and unsaturated optical amplifiers is negative for saturated amplifiers [Watanabe]. It cancels the positive chirp-parameter of modulator, causing asymmetrical optical power between the sidebands [Lee] and the optical amplification causes RF signal gain [Marti]. However the SOA adds noticeable noise to the system.

3.4.2 Calculation and simulation results

The frequency transfer function of the optical link:

$$H_{SOA+link}(f) = cos\left(\frac{\lambda^2 \cdot D \cdot \pi \cdot f^2 \cdot L}{c}\right) - LEF \cdot sin\left(\frac{\lambda^2 \cdot D \cdot \pi \cdot f^2 \cdot L}{c}\right) + \\ + j \cdot LEF \cdot \frac{f_c}{f} \cdot sin\left(\frac{\lambda^2 \cdot D \cdot \pi \cdot f^2 \cdot L}{c}\right) \tag{13}$$

Where D is the fibre dispersion parameter, L is the fibre length, f is the modulation frequency, c is the speed of light in vacuum, λ is the operating wavelength. For simplicity the linear loss and delay of the fibre are neglected.

The calculated RF responses of 400 km fibre for different chirp parameters of the optical transmitter are depicted in Fig.19. All the results have been calculated for an optical input power of 0 dBm in order to reduce the influence of the nonlinear effects in the fibre. By comparing the results to the reference case of a zero-chirp situation (*LEF*=0), for *LEF*<0, the achievable bandwidth increases.

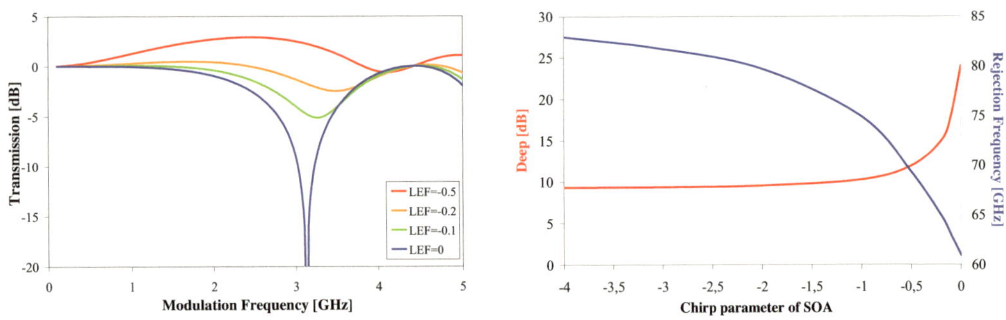

Fig. 19. Calculated quality of MW/MMW transmission for different SOA and system parameters, Optical wavelength=1550nm, D=16.8 ps/(nm·km)
(a) Transfer function with different chirp parameter of SOA, fiber length=100km
(b) Rejection frequency and the deep versus chirp parameter of SOA, fiber length=1km

In real radio over fibre systems the radio frequency carrier is modulated by digital information. The above presented dispersion compensation technique affects the eye diagram, the BER (Bit Error Rate) and the EVM (Error Vector Magnitude), which are the important quality factors of signal transmission. If the subcarrier frequency is close to one of the frequency notches caused by dispersion the eye diagram closes, the BER and EVM decay, the communication deteriorates or lost. Applying the SOA compensator the eye diagram opens, low BER is obtained and EVM can greatly benefit from the improved link performance. The simulations were executed with different modulation schemes on radio-over-fibre systems and VPI software was applied. Fig. 20 presents the quality of 4 QAM (quadrature amplitude modulation) radio signal transmission with different SOA chirp parameter, it shows the Symbol Error rate improvement applying SOA dispersion compensator. The EVM and SER increase versus chirp parameter. If the subcarrier frequency is close to the frequency notch, the communication lost. Applying a SOA dispersion compensator the constellation diagram can be detected, but the diagram rotates because of the additional phase modulation generating the SOA chirp.

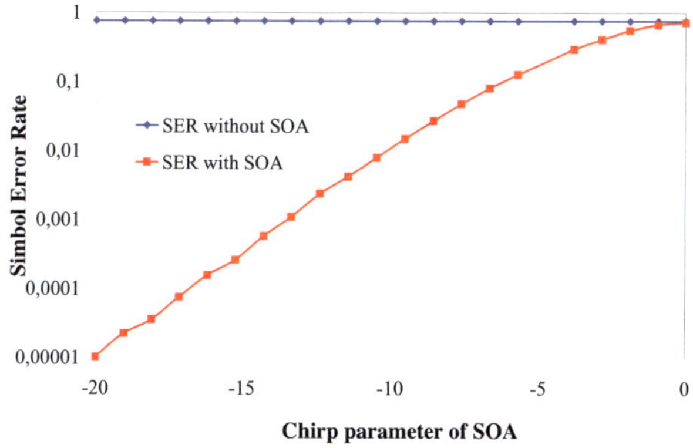

Fig. 20. Simulated Symbol Error Rate at frequency notch versus chirp parameter of SOA, L=500m, fsubcarrier=60GHz, modulation= 2048Mbps, 4QAM, without and with SOA

3.4.3 Experimental work

The calculated and simulated results can be verified by experimental work. The SOA under test was driven by different bias (dc) currents. The polarization state of the incoming optical power was optimized by polarization controller. The harmful effect of the optical reflection was eliminated by optical isolators. The required optical power and wavelength were produced by a tuneable laser source. The intensity modulated optical signal was detected by a photodetector. The setup was controlled by a computer program, hence the measurement parameters were carefully set by the program and the measurement results were processed and stored.

The measured RF responses of 50km fibre for different operation points of SOA are depicted in Fig. 21. By comparing the results to the reference case of a zero-chirp situation (*LEF*=0), for *LEF*<0, the achievable bandwidth increases. As the SOA bias current (optical gain) increases the frequency notches of the RF response are reduced and shifted to higher modulation frequencies. Based on the results, we may conclude that the interplay of chirp generated by the saturated SOA and chromatic dispersion enables a significant reduction of the dispersion-induced effect.

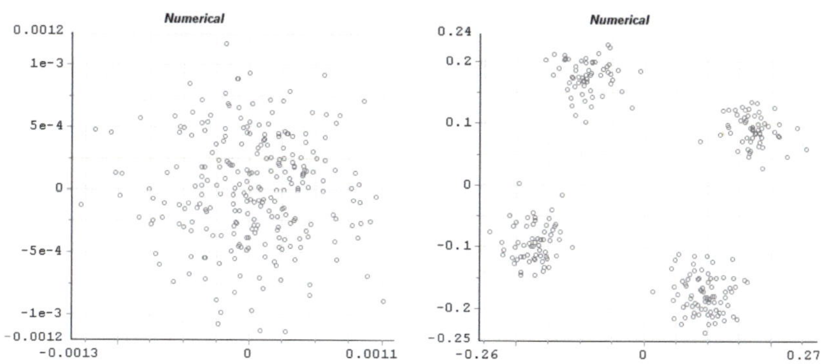

Fig. 21. Simulated constellation diagrams (a) without SOA, (b) with SOA compensator, L=500m, fsubcarrier=60GHz, modulation= 2048Mbps, 4QAM

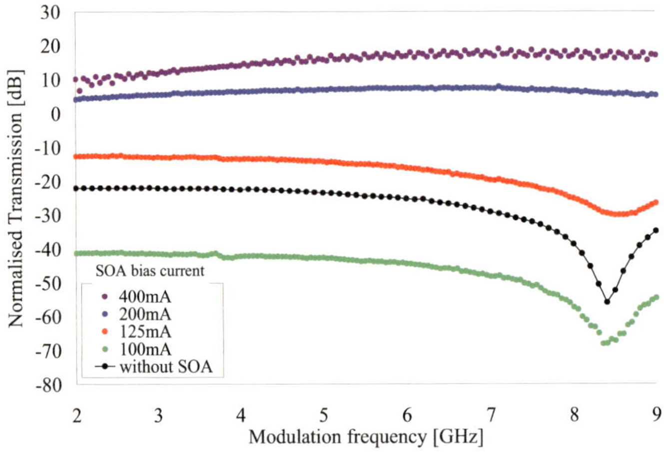

Fig. 22. Measured RF responses of the optical link for different SOA parameters
Length of the optical fiber=50km, Length of the reference optical fiber=4m, Optical wavelength=1550nm, bias current of the SOA=400mA, 200ma, 125mA, 100mA, Optical Gain=15dB, 13.5dB, 4,5dB, -9.5dB

4. Conclusion

This book chapter reviews the application of multifunctional Semiconductor Optical Amplifier in microwave photonics optical communication links. The device shows proper performance during the theoretical and experimental study. The noise, the linearity, the bandwidth and the dispersion effect of the multifunctional SOA applied in an optical link are studied. SOAs are potential candidates for modulation and detection tasks in RoF systems as their capabilities for detection/modulation, and amplification at the same time. So, cost effective designs are possible only employing one device at the subscriber's unit. On the other hand, the multifunctional SOA can improve the performance of microwave photonic links thanks to its dispersion compensation function.

5. References

A. J. Seeds: Broadband Fibre-Radio Access Networks, in Proc. MWP 1998, Princeton, New Jersey, USA, October 1998, pp.1-4

A. Ng'oma: Radio-over-Fibre Technology for Broadband Wireless Communication Systems, Eindhoven: Technische Universiteit Eindhoven, 2005. Proefschrift. - ISBN 90-386-1723-2

C. Arellano, J. Prat: Semiconductor Optical Amplifiers in Access Networks, in Proc. ICTON 2005, Barcelona, Spain, June 2005, paper WeA1.4

E. Udvary, T. Berceli: Linearity and chirp investigations on SOA as an external modulator in SCM systems, Proc. of the EUMA, Issue on MWP, vol. 3. pp. 217-222, Sept. 2007

E. Udvary, T. Berceli: Optical Subcarrier Label Swapping by Semiconductor Optical Amplifiers, J. Lightwave Technol., Special Issue on MWP, vol. 21. pp. 3221-3225, Dec. 2003

E. Udvary, T. Berceli: Semiconductor Optical Amplifier for Detection Function in Radio over Fiber Systems, *J. Lightwave Technol.*, Special Issue on MWP, vol. 26. pp. 2563-2570, Aug. 2008

Govind P. Agrawal: Fiber-Optic Communication Systems, John Wiley & Sons, 1985.

H.Ghafouri-Shiraz: "Laser Diode Amplifiers", John and Wiley, Chichester, 1996

Jesper Mork et al, "The modulation response of a Semiconductor Laser Amplifier". IEEE J. Sel. Topics. Quantum Electron., vol.5, no.3, pp.851-860, May/June, 1999.

J. Marti, F. Ramos, J. Herrera, "Experimental reduction of dispersion-induced effects in microwave optical links employing SOA boosters", Photonic Technology Letters, Vol. 13, No. 9, Sept. 2001, pp.999-1001

J-M. Kang, S-K. Han: A Hybrid WDM/SCM-PON Sharing Wavelength for Up- and Down-Link Using Reflective Semiconductor Optical Amplifier, Photon. Technol. Letters, Vol. 18, pp. 502-504, Feb. 2006

Josep Prat, et al.: Optical Network Unit Based on a Bidirectional Reflective Semiconductor Optical Amplifier for Fiber-to-the-Home Networks, Photon. Technology Letters, vol. 17. Jan. 2005

L. Occhi, L. Schares, G. Guekos, "Phase modeling based on the α factor in bulk semiconductor optical amplifiers", IEEE Journal of Selected Topics in Quantum Electronics, 2003, pp. 788-797

M.Connelly, "Wideband Semiconductor Optical Amplifier Steady-State Numerical Model", IEEE J. Quantum Electron., vol.37, no.3, pp.439-447, March, 2001.

M.Gustavsson, A.Karlsson, L.Thylen: "Traveling wave semiconductor laser amplifier detectors" J. of Lightwave Technology, vol.8., pp.610-617,1990

R.Olshansky et al, "Subcarrier Multiplexed Lightwave Systems for Broadband Distribution", J. Lightw. Technol., vol.7, no.9, pp.1329-1342, Sept. 1989.

Sang-Yun Lee, et al., "Reduction of chromatic dispersion effects and linearization of dual-drive Mach-Zehnder Modulator by using semiconductor optical amplifier in analog optical links" in Proc. ECOC 2002, September 8-12, 2002, Copenhagen, Denmark

T. Berceli, E. Udvary: Transmission Challenges of Cascaded Semiconductor Optical Amplifiers, in Proc. MWP 2005, Seoul, Korea, 12-14 October 2005, pp. 129-132

T. Watanabe et al., "Transmission performance of chirp-controlled signal by using semiconductor optical amplifier", IEEE Journal of Lightwave Technology, August 2000, pp. 1069-1077

T.Marozsák, "Transmission Characteristics of All Semiconductor Fiber OpticLinks Carrying Microwave Channels", European Microwave Conf., Paris, France, 2000, vol.2, pp.52-55.

6

Red Tunable High-Power Narrow-Spectrum External-Cavity Diode Laser Based on Tapered Amplifier

Mingjun Chi[1], Ole Bjarlin Jensen[1], Götz Erbert[2], Bernd Sumpf[2] and Paul Michael Petersen[1]
[1]*Department of Photonics Engineering, Technical University of Denmark,*
[2]*Ferdinand-Braun-Institut, Leibniz-Institut für Höchstfrequenztechnik,*
[1]*Denmark*
[2]*Germany*

1. Introduction

Diffraction-limited high-power narrow-spectrum red diode lasers are attractive for many applications, such as photodynamic therapy, laser display, and as a pump source to generate UV light by second harmonic generation (SHG). High-power, diffraction-limited diode lasers can be realized by the technology of lasers with a tapered gain-region (Kintzer et al., 1993; Donnelly et al., 1998; Wenzel et al., 2003; Paschke et al., 2005; Sumpf et al., 2009; Fiebig et al., 2010). The tapered laser devices can be used in applications where narrow-spectrum is not needed such as photodynamic therapy, but for other applications such as a pump source for UV light generation, the spectral quality of these devices has to be improved.

In order to improve the spectral quality of a tapered laser, different techniques are applied, such as a monolithically integrated master oscillator power amplifier by forming Bragg gratings in the semiconductor material (O'Brien et al., 1993, 1997a), injection locking to an external single-mode laser (Goldberg et al., 1993; Mehuys et al., 1993b; O'Brien et al., 1997b; Wilson et al., 1998; Ferrari et al., 1999; Spieβberger et al., 2011), and different external-cavity feedback techniques (Jones et al., 1995; Cornwell & Thomas, 1997; Morgott et al., 1998; Goyal et al., 1998; Pedersen & Hansen, 2005; Chi et al., 2005; Lucas-Leclin et al., 2008; Tien et al., 2008; Sakai et al., 2009). Up to 1 W output power at 668 nm from a Fabry-Perot tapered diode laser was obtained with a beam quality factor of 1.7, and the spectral width was smaller than 0.2 nm (Sumpf et al., 2007). Around 670 nm, tunable narrow-linewidth diffraction-limited output was also achieved from an injection-locking tapered diode laser system seeded with a single-mode external-cavity diode laser (Häring et al., 2007); the output power was up to 970 mW. A 670 nm micro-external-cavity tapered diode laser system was demonstrated with a reflecting volume Bragg grating as a feedback element; in continuous wave (CW) mode, more than 0.5 W output power was obtained, and in pulse mode, 5 W peak power was obtained with a beam quality factor of 10 and a spectral width below 150 pm (Tien et al., 2008). Up to 1.2 W output power at 675 nm from a tapered laser was obtained with a beam quality factor less than 1.3, the maximum conversion efficiency of 31% was reached at an output power of 1 W (Sumpf et al., 2011). External-cavity feedback based on a bulk diffraction grating in the Littrow configuration is a useful technique to achieve a tunable narrow-spectrum, high-power, diffraction-limited tapered diode laser system (Mehuys et al., 1993a; Jones et al., 1995; Goyal et al., 1997; Morgott et al., 1998; Chi et al., 2005). We have demonstrated such a tapered diode laser system around 668 nm with output power up to 810 mW; a beam quality factor of 3.4 was obtained with an output power of 600 mW (Chi et al., 2009).

In this chapter, three red tunable high-power narrow-spectrum diode laser systems based on three different tapered semiconductor optical amplifiers in Littrow external-cavity are demonstrated. Tapered device A is a 668 nm 2-mm-long tapered amplifier with a 0.5-mm-long index-guided ridge-waveguide section. Both tapered device B and C are 675 nm 2-mm-long tapered amplifier, the lengths of ridge-waveguide section are 0.5 mm for device B, and 0.75 mm for device C, respectively. The epitaxial structruce and the geometry of these tapered devices are described, and the data on the gain media of the devices are presented and compared.

Laser system A based on device A is tunable over a range of 16 nm centered at 668 nm. As high as 1.38 W output power is obtained at 668.35 nm. The emission spectral bandwidth is less than 0.07 nm throughout the tuning range, and the beam quality factor M^2 is 2.0 with an output power of 1.27 W.

Laser system B based on device B is tunable from 663 to 684 nm with output power higher than 0.55 W in the tuning range, as high as 1.25 W output power is obtained at 675.34 nm. The emission spectral bandwidth is less than 0.05 nm throughout the tuning range, and the beam quality factor M^2 is 2.07 at an output power of 1.0 W. Laser system C based on device C is tunable from 666 to 685 nm. As high as 1.05 W output power is obtained around 675.67 nm. The emission spectral bandwidth is less than 0.07 nm throughout the tuning range, and the beam quality factor M^2 is 1.13 at an output power of 0.93 W.

The properties of the three tapered diode laser systems are summarized and compared. As an example of application, Laser system C is used as a pump source for the generation of 337.6 nm UV light by single-pass frequency doubling in a bismuth triborate (BIBO) crystal. An output power of 109 μW UV light, corresponding to a conversion efficiency of 0.026%W^{-1} is attained.

2. Description of the gain media and the tapered devices

Tapered diode amplifiers (TDAs) consist of an index-guided ridge-waveguide section and a gain-guided tapered section (Kintzer et al., 1993; Sumpf et al., 2009). The ridge-waveguide section works as a spatial filter, thus diffraction-limited laser beam is available from tapered diode amplifiers, and high output power can be obtained due to the amplification from the gain-guided flared section.

The tapered amplifiers used in the experiment were grown using metal organic vapor phase epitaxy. The epitaxial structure of these tapered diode amplifiers were the same. As active layer, a 5-nm-thick compressively strained single $In_{1-x}Ga_xP$ quantum well was used for all the devices. The gallium content x was selected for an emission wavelength in the range between 670 and 680 nm. The single quantum well was embedded in the 500 nm thick $Al_{0.36}Ga_{0.16}InP$ *p*- and *n*-waveguide layers. For the 800 nm *n*-cladding layer $Al_{0.52}In_{0.48}P$ was

used, the 1000 nm *p*-cladding layer was made of $Al_{0.85}Ga_{0.15}As$, which allowed carbon doping with concentrations in the range of some 10^{18} cm^{-3} and a standard AlGaAs process. These epitaxial structures were also used for the manufacturing of tapered lasers as described previously (Sumpf et al., 2007, 2011).
Here we should mention that two factors influence the wavelength, i.e., the spectrum, of a tapered amplifier: the composition of the materials of the quantum well (in the red tapered amplifier, the gallium content x in the $In_{1-x}Ga_xP$ quantum well) and the strain between quantum well and waveguide. The detailed design on the wavelength of a tapered device is based on the semiconductor physics on quantum well, and this is out of the scope of this book chapter.

	Gain medium A	Gain medium B
λ / nm	668	675
Θ_{vert} / ° (FWHM)	31	30
Θ_{vert} / ° (95% power)	50	52
I_{th} / mA	330	315
η_D	0.66	0.70
T_0 / K	110	120
α_i / cm^{-1}	(1.8 ± 0.1)	(1.2 ± 0.1)
η_i	(0.90 ± 0.08)	(0.75 ± 0.02)
$\Gamma \cdot g_0$ / cm^{-1}	(19.6 ± 0.4)	(19.8 ± 0.4)

Table 1. Summary of the data for the gain media used in this study.

Tapered device A was made of gain medium A, and tapered device B and C were made of gain medium B. The gain media data were measured for uncoated broad-area devices (BADs) with a cavity length of 1 mm and a stripe width of 100 μm in pulsed mode. The vertical far field angles Θ_{vert} for the devices were about 30° (FWHM) and between 50° and 52° (95% power content). This relatively small vertical far field angle allows the use of standard optics with a moderate numerical aperture to collimate the output beam. The power-current characteristics and the spectra were measured for these BADs, and the threshold current I_{th}, the differential efficiency η_D, and the characteristic temperature of the threshold current T_0 were given in table 1. The threshold currents of gain medium A and B were 330 and 315 mA, respectively. The differential efficiency for gain medium A was slightly smaller compared to gain medium B. The characteristic temperatures of the threshold current were 110 and 120 K for gain medium A and B, respectively.
Assuming a logarithmic dependence of the gain on the current density, from the length-dependence measurement of threshold current density j_{th} and slope efficiency S, the gain medium data were obtained and given in Table 1. It showed that for medium A the internal loss α_i = 1.8 cm^{-1} was larger in comparison to medium B with α_i = 1.2 cm^{-1}. The internal efficiency for medium A was with η_i = 0.90 larger than η_i = 0.75 for medium B. The modal gain coefficient $\Gamma \cdot g_0$ remained constant for both gain media within the experimental uncertainty.
Based on these gain media, tapered diode amplifiers were processed with total cavity length of 2 mm. The straight index-guided ridge-waveguide section manufactured by reactive ion etching had a length of 0.5 mm for tapered device A and B, and 0.75 mm for tapered device

C. The width of the ridge-waveguide section was 7.5 μm for all the three tapered devices. The tapered gain-guided section was defined by ion implantation, and had a length of 1.5 mm for device A and B, and 1.25 mm for device C. The flared angle was 4° for all the three devices, and the output apertures for tapered amplifier A, B and C were 110, 112 and 95 μm, respectively.

The tapered amplifier facets were passivated and antireflection coated to achieve mirror reflectivities of 1% for the front facet, and 5×10^{-4} for the rear facet, respectively. The tapered devices were mounted *p*-side down on copper tungsten (CuW) or chemical vapour deposited (CVD) diamond submounts using AuSn solder. These subassemblies were mounted on standard C-mounts.

3. External-cavity tapered diode laser systems

In this section, the tunable external-cavity diode laser systems based on the three tapered amplifiers described in section 2 will be demonstrated.

3.1 External-cavity tapered diode laser system at 668 nm

The external-cavity configuration employed for diode laser system A is depicted in Fig. 1 (Chi et al., 2010). An aspherical lens of 3.1 mm focal length with a numerical aperture (NA) of 0.68 is used to collimate the beam from the back facet in both fast and slow axes. The bulk grating is ruled with 1200 grooves/mm and has a blazed wavelength of 750 nm. The grating is mounted in the Littrow configuration and oriented with the lines in the grating parallel to the active region of the amplifier in order to obtain optimum spectral filtering by the narrow aperture of the tapered device in the fast axis direction. A second aspherical lens of 3.1 mm focal length with an NA of 0.68 is used to collimate the beam from the output facet in the fast axis. Together with a cylindrical lens of 60 mm focal length, these two lenses collimate the output beam in the slow axis and compensate the astigmatism simultaneously. All the lenses are antireflection coated for the red wavelength. A beam splitter behind the cylindrical lens is used to reflect part of the output beam of the tapered diode laser system as the diagnostic beam, the spectral width and beam quality factor M^2 are measured in this beam. The output power of the laser system is measured behind the second aspherical lens.

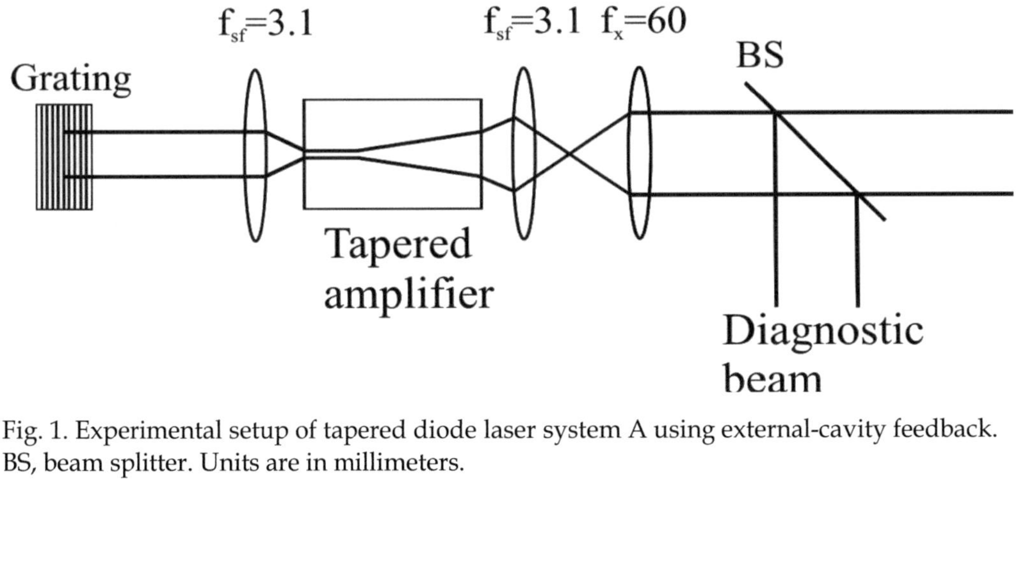

Fig. 1. Experimental setup of tapered diode laser system A using external-cavity feedback. BS, beam splitter. Units are in millimeters.

The grating is mounted in the Littrow configuration, this means the first-order diffraction beam of the grating propagates back towards the tapered amplifier. Therefore the laser cavity is formed between the diffraction grating and the front facet of the tapered amplifier. The tapered amplifier works as a gain medium in the laser cavity. When the injected current in the tapered device is higher than the threshold, the laser beam oscillates in the laser cavity. The emission wavelength of the laser system is tuned widely by rotating the diffraction grating because of the broad gain bandwidth (a few tens nanometers) of the tapered device. The emission spectrum of the laser system is narrowed significantly compared with the freely running tapered lasers (from a few hundreds picometers to a few tens picometers) due to the dispersion of the diffraction grating and the narrow aperture of the tapered device in the fast axis direction (Sumpf et al., 2007; Chi et al., 2010, 2011).
The laser is TE-polarized, i.e., linearly polarized along the slow axis. The temperature of the amplifier is controlled with a Peltier element and the amplifier is operated at 15 °C in the experiment. The tapered device lases without the grating feedback.
The power/current characteristics for laser system A with and without the external-cavity feedback are shown in Fig. 2. Without feedback, the threshold current is around 0.7 A, the slope efficiency is 0.63 W/A, the emission wavelength is around 667.1 nm, the roll-over takes place around 1.5 A, and an output power of 0.65 W is achieved with an injected current of 2.0 A. With the external-cavity feedback, the maximum power is obtained at a wavelength around 668.4 nm, the threshold current of the laser system is decreased to 0.5 A, the slope efficiency is increased to 1.05 W/A, the roll-over takes place around 1.7 A, and an output power of 1.38 W is obtained with an operating current of 2.0 A.

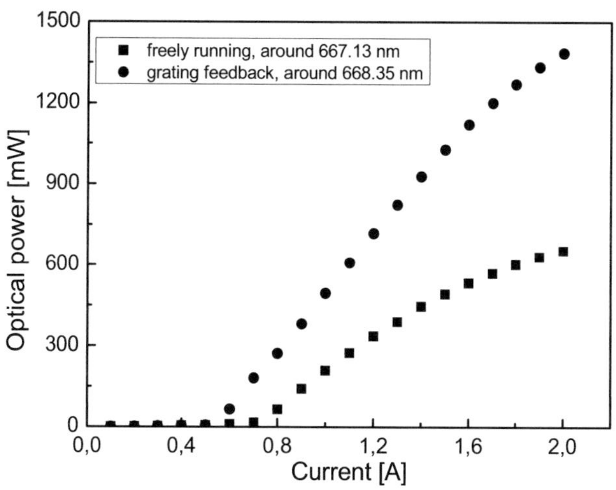

Fig. 2. The power/current characteristics for tapered diode laser system A with and without external-cavity feedback.

The output power at different wavelengths is shown in Fig. 3 at an operating current of 1.8 A. A maximum output power of 1.27 W is obtained at the wavelength of 668.38 nm, the output power is higher than 0.8 W in the tuning range from 659 to 675 nm. The tunable range is narrower compared with that in Ref. (Chi et al., 2009), since if we rotate the diffraction grating further to tune the wavelength, the laser system will jump to the freely running mode.

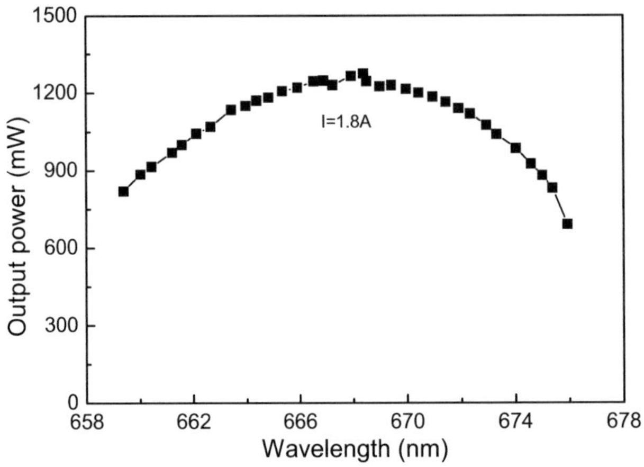

Fig. 3. Tuning curve of tapered diode laser system A at an operating current of 1.8 A.

The spatial beam quality of the output beam along the slow axis is determined by measuring the beam quality factor M^2 for the tunable external-cavity diode laser system. A spherical lens with a 100 mm focal length is used to focus the diagnostic beam. Then the beam width, W ($1/e^2$), is measured at various recorded positions Z along the optical axis - on both sides of the beam waist. The value of M^2 is obtained by fitting the measured data with the formula (Siegman & Townsend, 1993; Chi et al., 2004):

$$W = \left[W_0^2 + \left(\frac{4\lambda M^2}{\pi W_0} \right)^2 (Z - Z_0)^2 \right]^{1/2} . \quad (1)$$

Here W_0 is the beam waist width, λ is the wavelength of the laser system and Z_0 is the beam waist position.

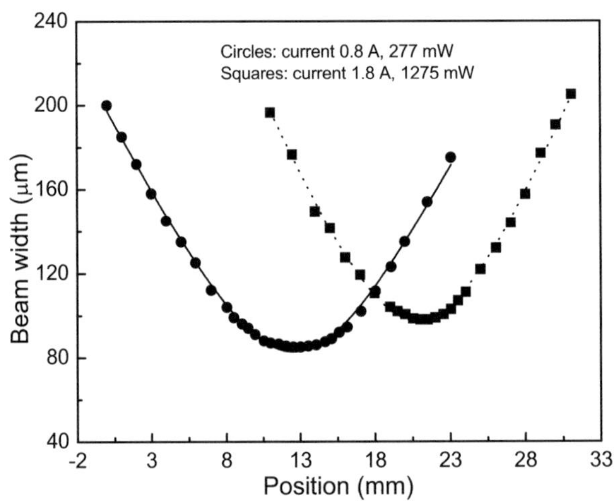

Fig. 4. Beam width of the output beam from tapered diode laser system A for the slow axis with an output power of 277 mW (circles and solid curve), and 1275 mW (squares and dotted curve). The curves represent the fits to the measured data using formula (1).

Figure 4 shows the measured beam widths of the output beam and the fitted curves using formula (1) with the output power of 277 and 1275 mW. The M^2 values are 1.39 ± 0.01 and 2.00 ± 0.01 with the output power of 277 and 1275 mW, respectively. For clarity, we have shifted the spatial position of the curves in the figure. In the experiments, the beam waists of the output beam with these two different output powers are located almost at the same position. This means that the astigmatism of the tapered device is rather stable over the whole power range.

The optical spectrum characteristic of the output beam from diode laser system A is measured using a spectrum analyzer (Advantest Corp. Q8347). A typical result measured at 667.91 nm with an output power of 1260 mW is shown in Fig. 5. The figure shows diode laser system A is operated in multiple longitudinal modes. The spectral bandwidth (FWHM) is 0.034 nm (the resolution of the spectrum analyzer is 3 pm), and the amplified spontaneous emission intensity is more than 20 dB suppressed. We find the spectral bandwidth of the output beam is less than 0.07 nm in the 16 nm tunable range.

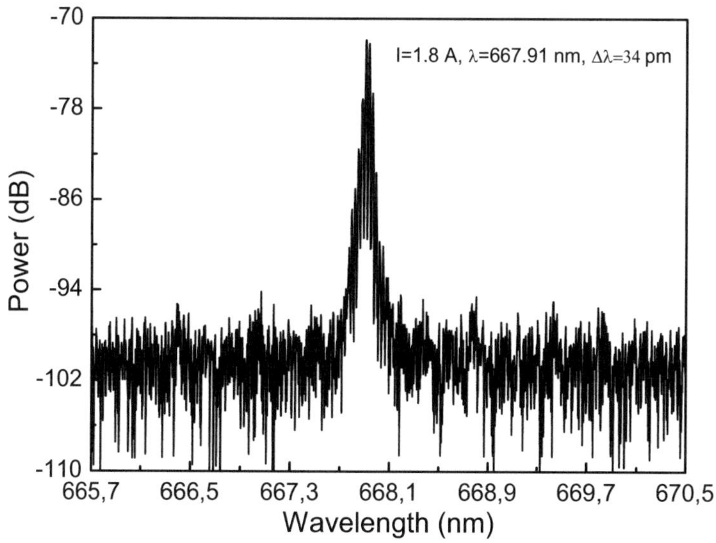

Fig. 5. The optical spectrum of the output beam from tapered diode laser system A with an output power of 1.26 W.

Compared with the results obtained previously in the similar wavelength range (Chi et al., 2009), the output power and spatial beam quality of the tapered diode laser system described above are improved significantly. The reason is the tapered semiconductor amplifier used here has improved properties. The length of the ridge-waveguide section of the 2 mm tapered device is 0.5 mm instead of 0.75 mm in the previous device. It has been verified that the beam quality of the 650 nm tapered diode lasers with shorter ridge-waveguide section is better compared with the device with longer ridge-waveguide section (Adamiec et al., 2009). This may be also the reason for better beam quality obtained here. The internal loss of the present device is less than 2.0 cm^{-1}, while the internal loss of the previous device is between 2.5 and 3.0 cm^{-1} (Sumpf et al., 2007). We believe that the lower internal loss is the reason for higher output power from the present external-cavity tapered diode laser system.

3.2 External-cavity tapered diode laser systems at 675 nm

The external-cavity configuration is the same for both tapered gain device B and C as depicted in Fig. 6 (Chi et al., 2011). The detailed description on the external-cavity tapered diode laser system can be found in section 3.1. An aspherical lens is used to collimate the beam from the back facet. A bulk grating is mounted in the Littrow configuration, and the laser cavity is formed between the diffraction grating and the front facet of the tapered amplifier. A second aspherical lens and a cylindrical lens are used to collimate the beam from the output facet. All the lenses are antireflection coated for the red wavelength. A beam splitter behind the cylindrical lens is used to reflect part of the output beam of the tapered diode laser system as the diagnostic beam, where the spectral bandwidth and the beam quality factor M^2 are measured. The output power of the laser system is measured behind the second aspherical lens.

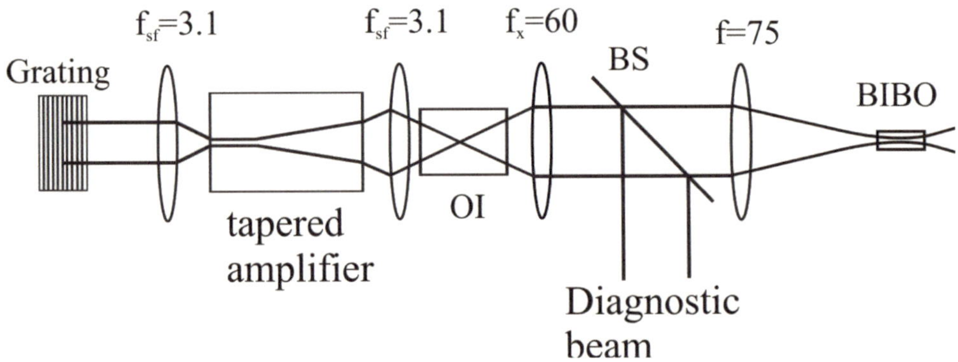

Fig. 6. Experimental setup of the external-cavity tapered diode laser system for SHG. BS, beam splitter; OI, optical isolator. Units are in millimeters.

The lasers are TE-polarized, i.e., linearly polarized along the slow axis. The temperature of the amplifiers is controlled with a Peltier element and the tapered amplifiers are operated at 20 °C in the experiment. The emission wavelength of the laser systems is tuned by rotating the diffraction grating.

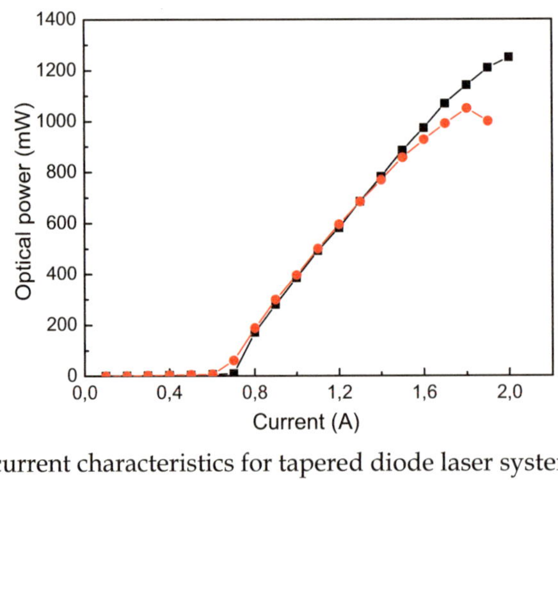

Fig. 7. The power/current characteristics for tapered diode laser system B (squares) and C (circles).

The power/current characteristics for these two tunable diode laser systems are shown in Fig. 7. For diode laser system B, the maximum output power is obtained at the wavelength 675.34 nm, the threshold current is around 0.7 A, the slope efficiency is 1.03 W/A, the roll-over takes place around 1.7 A, and an output power of 1.25 W is achieved with an injected current of 2.0 A. For diode laser system C, the output power is measured at the wavelength of 675.67 nm, the threshold current of the laser system is around 0.6 A, the slope efficiency is 0.99 W/A, the roll-over takes place around 1.4 A, an output power of 1.05 W is obtained with an operating current of 1.8 A, and the output power decreases with higher injected current.

The output power at different wavelengths for these two laser systems is shown in Fig. 8. For laser system B, at an operating current of 2.0 A, a maximum output power of 1250 mW is obtained at the wavelength of 675.34 nm. The output power is higher than 550 mW in the tuning range from 663 to 684 nm. For laser system C, a maximum output power of 1055 mW is obtained at the wavelength of 675.67 nm; the laser system is tunable from 666 to 685 nm with output power higher than 460 mW.

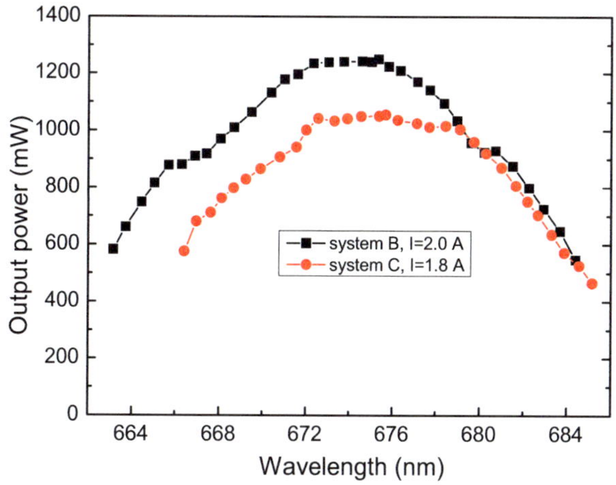

Fig. 8. Tuning curves of the tapered diode laser system B (squares) at an operating current of 2.0 A and system C (circles) at an operating current of 1.8 A.

For both diode laser system B and C, the spatial beam quality of the output beam along the fast axis is assumed to be diffraction-limited because of the waveguide structure of the tapered gain devices. The spatial beam quality of the output beam along the slow axis is determined by measuring the beam quality factor M^2. A spherical lens with a 65 mm focal length is used to focus the diagnostic beam. Then the beam width, W $(1/e^2)$, is measured at various recorded positions along the optical axis - on both sides of the beam waist. The value of M^2 is obtained by fitting the measured data with formula (1). Figure 9 shows the measured beam widths and the fitted curves for tapered diode laser system B at output powers of 385 and 1000 mW, where the M^2 values are 1.28 ± 0.02 and 2.07 ± 0.02, respectively. Figure 10 shows the measured beam widths and the fitted curves for tapered diode laser system C with the output power of 390 and 930 mW. The M^2 values are 1.20 ± 0.02 and 1.13 ± 0.02 at output powers of 390 and 930 mW, respectively. For clarity, we have shifted the spatial position of the curves in the figures. In the experiments, the waists of the output beam with different output powers are located almost at the same position. This indicates that the change of the astigmatism of the tapered device with different injection currents is negligible.

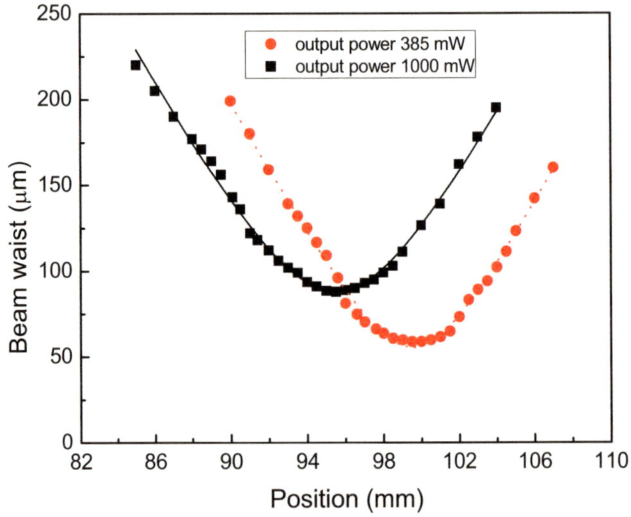

Fig. 9. Beam width of the output beam for the slow axis from diode laser system B with the output power of 385 mW (circles and dotted curve) and 1000 mW (squares and solid curve). The curves represent the fits to the measured data using formula (1).

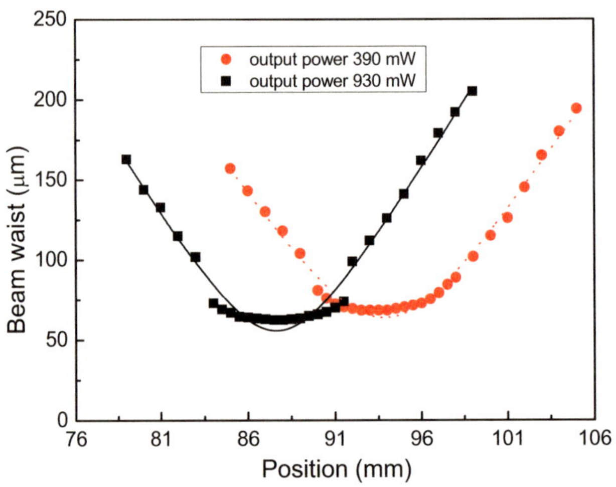

Fig. 10. Beam width of the output beam for the slow axis from diode laser system C with the output power of 390 mW (circles and dotted curve) and 930 mW (squares and solid curve). The curves represent the fits to the measured data using formula (1).

The optical spectrum characteristic of the output beam from diode laser system B and C is measured using a spectrum analyzer (Advantest Corp. Q8347). A typical result measured for laser system C at 675.04 nm with an output power of 930 mW is shown in Fig. 11. The figure shows that the diode laser system is operated in multiple longitudinal modes. The spectral bandwidth (FWHM) is 0.038 nm (the resolution of the spectrum analyzer is 3 pm), and the amplified spontaneous emission intensity is more than 20 dB suppressed. We find the spectral bandwidths of the output beams for diode laser system B and C are less than 0.05 and 0.07 nm throughout their tunable ranges, respectively.

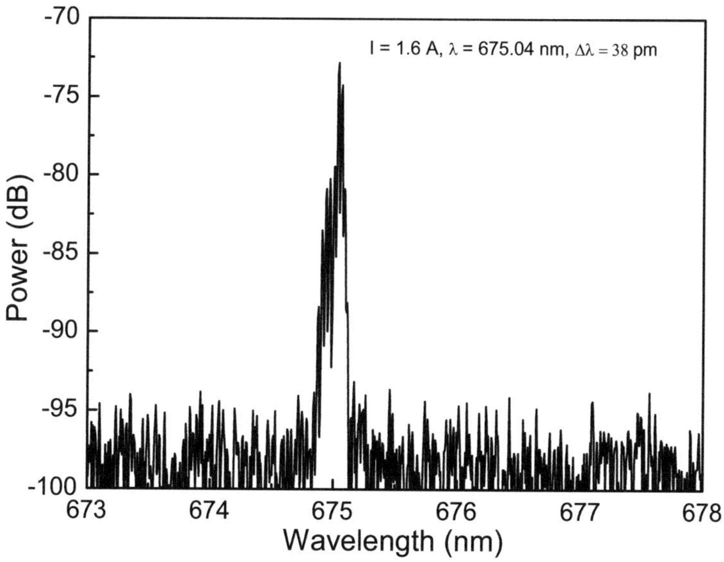

Fig. 11. Optical spectrum of the output beam from tapered diode laser system C with an output power of 930 mW.

Two tunable high-power 675 nm external-cavity diode laser systems based on tapered semiconductor optical amplifiers are demonstrated in this subsection. The main parameters of these two diode laser systems are summarized in Table 2; the results of diode laser system A described in subsection 3.1 are also listed in the table for comparison. Diode laser system B can produce more output power than diode laser system C, but the spatial beam quality of diode laser system C along the slow axis is significantly better than that of diode laser system B, especially with high output power. The two tapered devices are from the same wafer, so the different behaviour of the two laser systems is originated from the different sizes of the ridge-waveguide section and the tapered section. Diode laser system B produces higher output power since device B has a longer tapered gain section, but this laser system has worse spatial beam quality due to insufficient filtering of the light in the shorter ridge-waveguide section.

Compared with the results from diode laser system A in subsection 3.1, much more output power around 675.5 nm is obtained from laser system B and C, i.e., 1.25 and 1.05 W vs. 0.83 W; and diode laser system A cannot reach wavelengths longer than 676 nm.

Furthermore, we compare the results from the three laser systems based on different tapered gain devices. This is important for us to choose tapered gain devices for different applications.

	Laser system		
Parameter	A	B	C
Max. power (W)	1.38	1.25	1.05
Wavelength with max power (nm)	668.35	675.34	675.67
Tunable range (nm)	659-675	663-684	666-685
M^2 value	2.00±0.01 (1.27 W)	2.07±0.02 (1.0 W)	1.13±0.02 (0.93 W)
Spectral bandwidth (nm)	<0.07	<0.05	<0.07

Table 2. Summary of the main parameters for diode laser system A, B and C.

4. UV light generation by SHG

UV light sources are interesting in many fields such as biophotonics/chemical photonics, material processing, and optical data storage. Although diode lasers based on AlGaN have been demonstrated around 340 nm in pulse mode recently (Yoshida et al., 2008a, 2008b), frequency doubling of a red laser beam through a nonlinear crystal is still an efficient method to generate CW light in this wavelength range (Mizuuchi & Yamamoto, 1996; Mizuuchi et al., 1997, 2003a, 2003b; Knappe et al., 1998). UV light around 340 nm has been achieved by single pass SHG in bulk periodically poled $LiTaO_3$ (Mizuuchi & Yamamoto, 1996; Mizuuchi et al., 1997) and $MgO:LiNbO_3$ (Mizuuchi et al., 2003a) crystals and also achieved in a periodically poled $MgO:LiNbO_3$ ridge waveguide (Mizuuchi et al., 2003b); but so far these periodically poled devices are not commercially available, and no UV light shorter than 340 nm has been demonstrated with these first-order periodically poled devices.

Here UV light around 337.5 nm will be generated using the external-cavity tapered diode laser system developed above as a pump source. The generated CW UV light source will be used as the excitation source for fluorescence diagnostics. Compared with other UV laser sources around 337 nm, such as a CW krypton-ion laser and a pulsed nitrogen laser, the UV laser source based on a tunable tapered diode laser system is far more simple, compact, and easy to operate.

A BIBO nonlinear crystal is used for frequency doubling of the 675.2 nm red light to the 337.6 nm UV light due to its relatively high effective nonlinear coefficient. The 10 mm long BIBO crystal with an aperture of 4 mm × 4 mm is cut with $\theta = 137.7^{\circ}$ and $\varphi = 90^{\circ}$ for type-I phase matching (*eeo*) and antireflection coated on both end surfaces for 675/337.5 nm. The spectral bandwidth of both laser system B and C is narrow enough for frequency doubling through the BIBO crystal (the acceptable spectral bandwidth of this crystal is around 0.2 nm). Laser system C is chosen as the pump source for the frequency doubling experiment due to its better spatial beam quality.

A 30 dB optical isolator is inserted between the aspherical lens and the cylindrical lens in the output beam to avoid feedback from the optical components and the nonlinear crystal, as shown in Fig. 6. A biconvex lens of 75 mm focal length is used to focus the red fundamental beam into the BIBO crystal. The available output power of the fundamental beam in front of the crystal is 650 mW. The size of the focus $w_s \times w_f$ is around 70 μm × 35 μm, where w_s and w_f are the beam waists (diameters at $1/e^2$) in the slow and fast axes, respectively. The elliptical beam is used to reduce the effects of walk-off in the BIBO crystal. The walk-off angle in our crystal is 72.9 mrad, corresponding to a heavy walk-off parameter B of 15.1. The elliptical beam was proved to be optimum in the experiments, in good agreement with the theory of frequency doubling using elliptical beams (Boyd & Kleinman, 1968; Steinbach et al., 1996). The slight change in astigmatism with output power will cause the focusing conditions to vary slightly at different power levels. In the experiments, the astigmatism was corrected at maximum pump power. Two dichroic beam splitters separate the fundamental beam from the second harmonic output beam.

The wavelength of the fundamental beam is tuned to 675.16 nm, and the temperature of the crystal is 19.8 °C. Figure 12 shows the measured second harmonic power as a function of fundamental power. The curve represents a quadratic fitting. A maximum of 109 μW UV light is obtained with a fundamental pump power of 650 mW. The conversion efficiency η is 0.026%W^{-1}, compared to a conversion efficiency of 0.019%W^{-1} for a single-pass frequency doubling through a 15 mm long $LiIO_3$ bulk crystal (Knappe et al., 1998), and the theoretically calculated value is 0.040%W^{-1} (Steinbach et al., 1996).

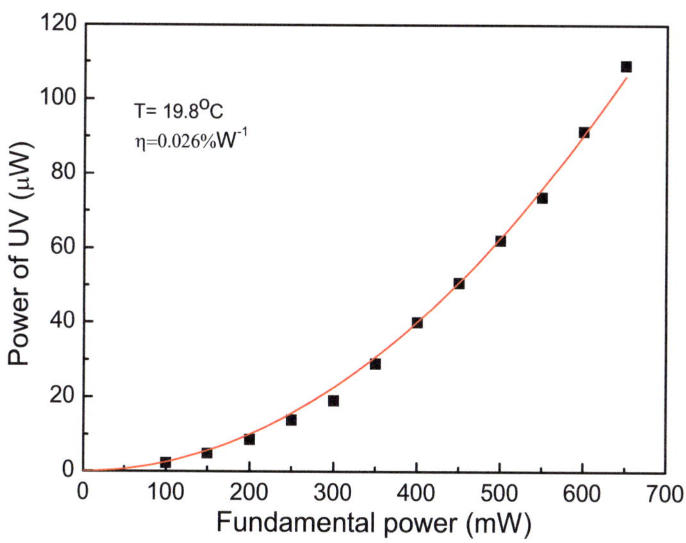

Fig. 12. Second harmonic power as a function of fundamental power. The squares are measured data, and the curve is a quadratic fit.

5. Conclusion

In conclusion, three tunable high-power narrow-spectrum red diode laser systems based on tapered semiconductor optical amplifiers in Littrow external-cavity are demonstrated. The epitaxial structruce and the geometry of these tapered devices are described, and the data on the gain media of the devices are presented and compared.

For diode laser system A at 668 nm, an output power of 1.38 W is obtained with an injected current of 2.0 A, and the laser system is tunable from 659 to 675 nm with output power over 800 mW. This is to our knowledge the highest output power from a tunable diode laser system in this wavelength range. The spectral bandwidth of the output beam is less than 0.07 nm.

Both diode laser system B and C are tunable in a 20 nm range centered at 675 nm, and the spectral bandwidth of the output beam for both diode laser systems is less than 0.07 nm in their tunable ranges. The maximum output power is 1.25 W obtained from laser system B at the wavelength of 675.3 nm, and the maximum output power from laser system C is 1.05 W obtained at the wavelength of 675.6 nm. The beam quality factor M^2 is 2.07 with the output power of 1.0 W for laser system B, and the M^2 value is 1.13 with the output power of 0.93 W for laser system C. Laser system C is used as the pump source for the generation of UV light by single-pass frequency doubling in a BIBO crystal. An output power of 109 μW UV light at 337.6 nm, corresponding to a conversion efficiency of 0.026%W^{-1} is attained.

6. Acknowledgment

The authors acknowledge the financial support of the European community through the project WWW.BRIGHTER.EU (grant No. FP6-IST-035266).

7. References

Adamiec, P.; Sumpf, B.; Rüdiger, I.; Fricke, J.; Hasler, K.-H.; Ressel, P.; Wenzel, H.; Zorn, M.; Erbert, G. & Tränkle, G. (2009). Tapered lasers emitting at 650 nm with 1 W output power with nearly diffraction-limited beam quality. *Opt. Lett.*, Vol. 34, No. 16, (August 2009), pp. 2456-2458, ISSN 0146-9592

Boyd, G.D. & Kleinman, D.A. (1968). Parametric interaction of focused Gaussian light beams. *J. Appl. Phys.*, Vol. 39, No. 8, (July 1968), pp. 3597-3639, ISSN 0021-8979

Chi, M.; Bøgh, N.-S.; Thestrup, B. & Petersen, P.M. (2004). Improvement of the beam quality of a broad-area diode laser using double feedback from two external mirrors. *Appl. Phys. Lett.*, Vol. 85, No. 7, (August 2004), pp. 1107-1109, ISSN 0003-6951

Chi, M.; Jensen, O.B.; Holm, J.; Pedersen, C.; Andersen, P.E.; Erbert, G.; Sumpf, B. & Petersen, P.M. (2005). Tunable high-power narrow-linewidth semiconductor laser based on an external-cavity tapered amplifier. *Opt. Express*, Vol. 13, No. 26, (December 2005), pp. 10589-10596, ISSN 1094-4087

Chi, M.; Jensen, O.B.; Erbert, G.; Sumpf, B. & Petersen, P.M. (2009). Tunable high-power narrow-linewidth semiconductor laser based on an external-cavity tapered amplifier at 670 nm. *Proceeding of the 8th Pacific Rim conference on Lasers and*

Electro-optics, WD1-4, ISBN 978-1-4244-3830-3, Shanghai, China, Aug. 30-Sep. 3, 2009

Chi, M.; Erbert, G.; Sumpf, B. & Petersen, P.M. (2010). Tunable high-power narrow-spectrum external-cavity diode laser based on tapered amplifier at 668 nm. *Opt. Lett.*, Vol. 35, No. 10, (May 2010), pp. 1545-1547, ISSN 0146-9592

Chi, M.; Jensen, O.B.; Erbert, G.; Sumpf, B. & Petersen, P.M. (2011). Tunable high-power narrow-spectrum external-cavity diode laser at 675 nm as a pump source for UV generation. *Appl. Opt.*, Vol. 50, No. 1, (January 2011), pp. 90-94, ISSN 0003-6935

Cornwell, D.M. & Thomas, H.J. (1997). High-power (>0.9 W cw) diffraction-limited semiconductor laser based on a fiber Bragg grating external cavity. *Appl. Phys. Lett.*, Vol. 70, No. 6, (February 1997), pp. 694-695, ISSN 0003-6951

Donnelly, J.P.; Walpole, J.N.; Groves, S.H.; Bailey, R.J.; Missaggia, L.J.; Napoleone, A.; Reeder, R.E. & Cook, C.C. (1998). 1.5-μm tapered-gain-region lasers with high-CW output powers. *IEEE Photon. Technol. Lett.*, Vol. 10, No. 10, (October 1998), pp. 1377-1379, ISSN 1041-1135

Ferrari, G.; Mewes, M.; Schreck, F. & Salomon, C. (1999). High-power multiple-frequency narrow-linewidth laser source based on a semiconductor tapered amplifier. *Opt. Lett.*, Vol. 24, No. 3, (February 1999), pp. 151-153, ISSN 0146-9592

Fiebig, C.; Eppich, B.; Paschke, K. & Erbert, G. (2010). High-brightness 980-nm tapered laser-optimization of the laser rear facet. *IEEE Photon. Technol. Lett.*, Vol. 22, No. 5, (March 2010), pp. 341-343, ISSN 1041-1135

Goldberg, L.; Mehuys, D.; Surette, M.R. & Hall, D.C. (1993). High-power, near-diffraction-limited large-area traveling-wave semiconductor amplifiers. *IEEE J. Quantum Electron.*, Vol. 29, No. 6, (June 1993), pp. 2028-2043, ISSN 0018-9197

Goyal, A.K.; Gavrilovic, P. & Po, H. (1997). Stable single-frequency operation of a high-power external cavity tapered diode laser at 780 nm. *Appl. Phys. Lett.*, Vol. 71, No. 10, (September 1997), pp. 1296-1298, ISSN 0003-6951

Goyal, A.K.; Gavrilovic, P. & Po, H. (1998). 1.35 W of stable single-frequency emission from an external-cavity tapered oscillator utilizing fiber Bragg grating feedback. *Appl. Phys. Lett.*, Vol. 73, No. 5, (August 1998), pp. 575-577, ISSN 0003-6951

Häring, R.; Sumpf, B.; Erbert, G.; Tränkle, G.; Lison, F. & Kaenders, W.G. (2007). 670 nm semiconductor lasers for Lithium spectroscopy with 1 W. *Proceedings of SPIE*, Vol. 6485, 648516, ISBN 978-0-8194-6598-6, San Jose, CA, USA, Jan. 20-25, 2007

Jones, R.J.; Gupta, S.; Jain, R.K. & Walpole, J.N. (1995). Near-diffraction-limited high power (~1 W) single longitudinal mode CW diode laser tunable from 960 to 980 nm. *Electron. Lett.*, Vol. 31, No. 19, (September 1995), pp. 1668-1669, ISSN 0013-5194

Kintzer, E.S.; Walpole, J.N.; Chinn, S.R.; Wang, C.A. & Missaggia, L.J. (1993). High-power, strained-layer amplifiers and lasers with tapered gain regions. *IEEE Photon. Technol. Lett.*, Vol. 5, No. 6, (June 1993), pp. 605-608, ISSN 1041-1135

Knappe, R.; Laue, C.K. & Wallenstein, R. (1998). Tunable UV-source based on frequency-doubled red diode laser oscillator-amplifier system. *Electron. Lett.*, Vol. 34, No. 12, (June 1998), pp. 1233-1234, ISSN 0013-5194

Lucas-Leclin, G.; Paboeuf, D.; Georges, P.; Holm, J.; Andersen, P.; Sumpf, B. & Erbert, G. (2008). Wavelength stabilization of extended-cavity tapered lasers with volume Bragg gratings. *Appl. Phys. B*, Vol. 91, No. 3-4, (June 2008), pp. 493-498, ISSN 1432-0649

Mehuys, D.; Welch, D. & Scifres, D. (1993a). 1 W CW, diffraction-limited, tunable external-cavity semiconductor laser. *Electron. Lett.*, Vol. 29, No. 14, (July 1993), pp. 1254-1255, ISSN 0013-5194

Mehuys, D.; Goldberg, L. & Welch, D.F. (1993b). 5.25-W CW near-diffraction-limited tapered –stripe semiconductor optical amplifier. *IEEE Photon. Technol. Lett.*, Vol. 5, No. 10, (October 1993), pp. 1179-1182, ISSN 1041-1135

Mizuuchi, K. & Yamamoto, K. (1996). Generation of 340-nm light by frequency doubling of a laser diode in bulk periodically poled $LiTaO_3$. *Opt. Lett.*, Vol. 21, No. 2, (January 1996), pp. 107-109, ISSN 0146-9592

Mizuuchi, K.; Yamamoto, K. & Kato, M. (1997). Generation of ultraviolet light by frequency doubling of a red laser diode in a first-order periodically poled bulk $LiTaO_3$. *Appl. Phys. Lett.*, Vol. 70, No. 10, (March 1997), pp. 1201-1203, ISSN 0003-6951

Mizuuchi, K.; Morikawa, A.; Sugita, T. & Yamamoto, K. (2003a). Efficient second-harmonic generation of 340-nm light in a 1.4-μm periodically poled bulk $MgO:LiNbO_3$. *Jpn. J. Appl. Phys.*, Vol. 42, No. 2A, (February 2003), pp. L90-L91, ISSN 0021-4922

Mizuuchi, K.; Sugita, T.; Yamamoto, K.; Kawaguchi, T.; Yoshino, T. & Imaeda, M. (2003b). Efficient 340-nm light generation by a ridge-type waveguide in a first-order periodically poled $MgO:LiNbO_3$. *Opt. Lett.*, Vol. 28, No. 15, (August 2003), pp. 1344-1346, ISSN 0146-9592

Morgott, S.; Chazan, P.; Mikulla, M.; Walther, M.; Kiefer, R.; Braunstein, J. & Weimann, G. (1998). High-power near-diffraction-limited external cavity laser, tunable from 1030 to 1085 nm. *Electron. Lett.*, vol. 34, No. 6, (March 1998), pp. 558-559, ISSN 0013-5194

O'Brien, S.; Welch, D.F.; Parke, R.A.; Mehuys, D.; Dzurko, K.; Lang, R.J.; Waarts, R. & Scifres, D. (1993). Operating characteristics of a high-power monolithically integrated flared amplifier master oscillator power amplifier. *IEEE J. Quantum Electron.*, Vol. 29, No. 6, (June 1993), pp. 2052-2057, ISSN 0018-9197

O'Brien, S.; Lang, R.; Parke, R.; Major, J.; Welch, D.F. & Mehuys, D. (1997a). 2.2-W continuous-wave diffraction-limited monolithically integrated master oscillator power amplifier at 854 nm. *IEEE Photon. Technol. Lett.*, Vol. 9, No. 4, (April 1997), pp. 440-442, ISSN 1041-1135

O'Brien, S.; Schoenfelder, A. & Lang, R.J. (1997b). 5-W CW diffraction-limited InGaAs broad-area flared amplifier at 970 nm. *IEEE Photon. Technol. Lett.*, Vol. 9, No. 9, (September 1997), pp. 1217-1219, ISSN 1041-1135

Paschke, K.; Sumpf, B.; Dittmar, F.; Erbert, G.; Staske, R.; Wenzel, H. & Tränkle, G. (2005). Nearly diffraction limited 980-nm tapered diode lasers with an output power of 7.7 W. *IEEE J. Sel. Top. Quantum Electron.*, Vol. 11, No. 5, (September/October 2005), pp. 1223-1227, ISSN 1077-260X

Pedersen, C. & Hansen, R.S. (2005). Single frequency, high power, tapered diode laser using phase-conjugated feedback. *Opt. Express*, Vol. 13, No. 11, (May 2005), pp. 3961-3968, ISSN 1094-4087

Sakai, K.; Itakura, S.; Shimada, N.; Shibata, K.; Hanamaki, Y.; Yagi, T. & Hirano, Y. (2009). High-power tapered unstable-resonator laser diode with a fiber-Bragg grating reflector. *IEEE Photon. Technol. Lett.*, Vol. 21, No. 16, (August 2009), pp. 1103-1105, ISSN 1041-1135

Siegman, A.E. & Townsend, S.W. (1993). Output beam propagation and beam quality from a multimode stable-cavity laser. *IEEE J. Quantum Electron.*, Vol. 29, No. 4, (April 1993), pp. 1212-1217, ISSN 0018-9197

Spieβberger, S.; Schiemangk, M.; Sahm, A.; Wicht, A.; Wenzel, H.; Peters, A.; Erbert, G. & Tränkle, G. (2011). Micro-integrated 1 Watt semiconductor laser system with a linewidth of 3.6 kHz. *Opt. Express*, Vol. 19, No. 8, (April 2011), pp. 7077-7083, ISSN 1094-4087

Steinbach, A.; Rauner, M.; Cruz, F.C. & Bergquist, J.C. (1996). CW second harmonic generation with elliptical Gaussian beams. *Opt. Commun.*, Vol. 123, No. 1-3, (January 1996), pp. 207-214, ISSN 0030-4018

Sumpf, B.; Erbert, G.; Fricke, J.; Froese, P.; Häring, R.; Kaenders, W.G.; Klehr, A.; Lison, F.; Ressel, P.; Wenzel, H.; Weyers, M.; Zorn, M. & Tränkle, G. (2007). 670 nm tapered lasers and amplifiers with output powers P ≥ 1 W and nearly diffraction limited beam quality. *Proceedings of SPIE,* Vol. 6485, 648517, ISBN 978-0-8194-6598-6, San Jose, CA, USA, Jan. 20-25, 2007

Sumpf, B.; Hasler, K.-H.; Adamiec, P.; Bugge, F.; Dittmar, F.; Fricke, J.; Wenzel, H.; Zorn, M.; Erbert, G. & Tränkle, G. (2009). High-Brightness quantum well tapered lasers. *IEEE J. Sel. Top. Quantum Electron.,* Vol. 15, No. 3, (May/June 2009), pp. 1009-1020, ISSN 1077-260X

Sumpf, B.; Adamiec, P.; Zorn, M.; Wenzel, H. & Erbert, G. (2011). Nearly Diffraction-limited tapered lasers at 675 nm with 1-W output power and conversion efficiencies above 30%. *IEEE Photon. Technol. Lett.*, Vol. 23, No. 4, (February 2011), pp. 266-268, ISSN 1041-1135

Tien, T.Q.; Maiwald, M.; Sumpf, B.; Erbert, G. & Tränkle, G. (2008). Microexternal cavity tapered lasers at 670 nm with 5 W peak power and nearly diffraction-limited beam quality. *Opt. Lett.,* Vol. 33, No. 22, (November 2008), pp. 2692-2694, ISSN 0146-9592

Wenzel, H.; Sumpf, B. & Erbert, G. (2003). High-brightness diode lasers. *Comptes Rendus Physique,* Vol. 4, No. 6, (July-August 2003), pp. 649-661, ISSN 1631-0705

Wilson, A.C.; Sharpe, J.C.; McKenzie, C.R.; Manson, P.J. & Warrington, D.M. (1998). Narrow-linewidth master-oscillator power amplifier based on a semiconductor tapered amplifer. *Appl. Opt.*, Vol. 37, No. 21, (July 1998), pp. 4871-4875, ISSN 0003-6935

Yoshida, H.; Yamashita, Y.; Kuwabara, M. & Kan, H. (2008a). A 342-nm ultraviolet AlGaN multiple-quantum-well laser diode. *Nat. Photon.*, Vol. 2, No. 9, (September 2008), pp. 551-554, ISSN 1749-4885

Yoshida, H.; Yamashita, Y.; Kuwabara, M. & Kan, H. (2008b). Demonstration of an ultraviolet 336 nm AlGaN multiple-quantum-well laser diode. *Appl. Phys. Lett.*, Vol. 93, No. 24, (December, 2008), 241106, ISSN 0003-6951

7

Doped Fiber Amplifier Characteristic Under Internal and External Perturbation

Siamak Emami[1], Hairul Azhar Abdul Rashid[2], Seyed Edris Mirnia[1], Arman Zarei[1], Sulaiman Wadi Harun[1] and Harith Ahmad[1]
[1]University of Malaya Malaysia,
[2]Multimedia University Malaysia,
Malaysia

1. Introduction

Significant effort has been made in recent years to improve the Doped Fiber amplifier gain and noise figure. Extend the optical bandwidth of doped fiber amplifiers beyond the traditional 1550nm band, making the excellent EDFA characteristics available in a wider spectral region also was the main effort in optical amplifier fields. Several techniques have been developed to improve gain and shift the gain to the shorter wavelength region. In this chapter, the effects of external perturbation such as macro-bending and fiber length and internal perturbation such as transversal distribution profile and doped concentration on doped fiber performance have been demonstrated (S.D.Emami et al., 2010).

A macro-bending approach is demonstrated to increase a gain and noise figure at a shorter wavelength region of EDFA. The conventional double-pass configuration is used for the EDFA to obtain a higher gain with a shorter length and lower pump power. The macro-bending suppresses the ASE at longer wavelength to achieve a higher population inversion at shorter wavelengths. Without the bending, the peak ASE at 1530nm, which is a few times higher than the ASE at the shorter wavelength, would deplete the population inversion and suppresses the gain in this region.

Macro-bending is introduced as a new method to increase gain flatness and bandwidth of EDFA in C-band region. Varying the bending radius and doped fiber length leads to the optimized condition with flatter and broader gain profile. Under the optimized condition, gain at shorter wavelengths is increased due to increment of population inversion which results in gain reduction in the longer wavelength regions. The balance of these two effects in the optimized condition has a significant result in achieving a flattened and broadened gain profile.

This technique is also capable to compensate the fluctuation in operating temperatures due to proportional temperature sensitivity of absorption cross section and bending loss of the aluminosilicate EDF . This new approach can be used to design a temperature insensitive EDFA for application in a real optical communication system which operates at different environments but still maintaining the gain characteristic regardless of temperature variations. The effect of macro-bending on high concentration EDFA using optimized

bending radius and length of the doped fiber is demonstrated. This gain increment compensates the gain reduction of the EDF before applying macro-bending and result in a flat and broad gain spectrum.
One of the many EDFA optimization parameters reported includes the Erbium Transversal Distribution Profile (TDP). The Erbium TDP is essential in determining the overlap factor, which affects the absorption and emission dynamics of the EDFA. At the end of this chapter, numerical models of different Erbium TDP is demonstrated and later verified by experiment. The model considers the overlap factor and absorption/ emission dynamics for different Erbium TDP. Results indicate a high performance EDFA is achievable with an optimized and yet realistic Erbium TDP.

2. Macrobending effects on doped fiber amplifier

In the first part of this chapter, a macrobending approach is demonstrated to increase the gain and noise figure at a shorter wavelength region of EDFA. The conventional double pass configuration is used for the EDFA to obtain a higher gain with a shorter length and lower pump power. The macrobending suppresses the ASE at a longer wavelength to achieve a higher population inversion at shorter wavelengths. Without the bending, the peak ASE at 1530 nm, which is a few times higher than the ASE at the shorter wavelength, would deplete the population inversion and suppresses the gain in this region (Harun et al., 2008).
The configuration of the EDFA is based on a standard double-pass configuration, where a circulator was used at the input and output ends of the EDF to couple light out of the amplifier and to allow the double propagation of light in the gain medium, respectively. The EDF is pumped by a 980-nm laser diode using a propagating pump scheme. The commercial EDF used is 15 m long with an erbium ion concentration of 440 ppm. A tunable laser source is used to characterize the amplifier in conjunction with an optical spectrum analyzer (OSA). The amplifier is characterized in the wavelength region between 1480 to 1560 nm in terms of the gain and noise figure under changes in the optical power. Before the amplifier experiment, the optical loss of the EDF was characterized for both cases with and without macrobending. The macrobending is obtained by winding the EDF in a bobbin with various radiuses between 0.35 and 0.50 mm (Daud, et al. 2008).
The optical losses of EDF were measured against wavelengths at various radius of macro-bending and the result is compared to the straight EDF. Then the bending loss spectrum (dB/m) is obtained by taking the difference of the optical loss measurement between bent and straight EDF. Fig. 1 shows the bending loss spectrum at various bending radius between 0.35 to 0.50 mm. The experimental result is in agreement with the earlier reported theoretical prediction on bending loss in optical fiber (Thyagarajan & Kakkar, 2004), which uses a simple infinite cladding model. The theoretical result shows that the bending loss profile is almost exponential with respect to wavelength, with strong dependencies on fiber bending radius and refractive index profile. Bending of optical fiber, including EDF causes the propagating power of the guided modes to be transferred into cladding, which in turn resulted in loss of power and therefore the bending loss spectrum is obtained as shown in Fig. 2. The bending loss has a strong spectral variation because of the proportional changes of the mode field diameter with the signal wavelength. At bending radius of 0.40 mm, the experimental result shows that the bending loss is drastically increase (>10 dB/m) at the wavelengths above 1505 nm whereas the minimal loss is observed at the wavelengths below 1505 nm. This provides the ASE suppression of more than 270 dB at 1530 nm, which allows

a higher attainable gain at a shorter wavelength region. This result shows that the distributed ASE filtering can be achieved by macro-bending of the fiber at an optimally chosen radius. This characteristic can be used in research of S-band EDFA and fiber lasers (Daud, et al. 2008).

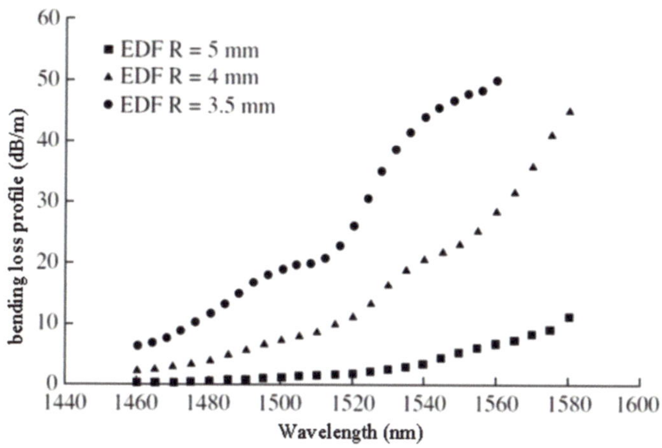

Fig. 1. EDF bending loss profile (dB/m) against wavelength (nm) for different bending radius (3.5 mm, 4 mm and 5 mm).

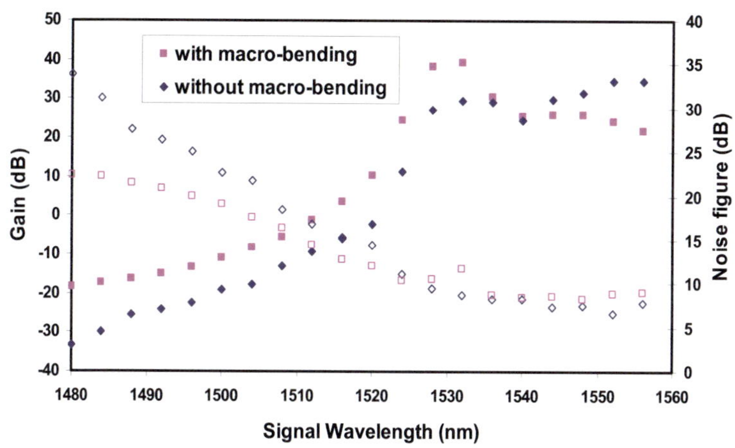

Fig. 2. Gain (solid symbols) and noise figure (hollow symbols) spectra with and without the macro-bending effect. The input signal and pump power is fixed at -30dBm and 100mW, respectively.

Fig. 2 shows the variation of gain and noise figure across the input signal wavelength for the double-pass EDFA with and without the macro-bending. The input signal and 980nm pump powers is fixed at -30 dBm and 100 mW respectively. The bending radius is set at 4 mm in case of the amplifier with the macro-bending. As shown in the figure, the gain enhancements of about 12 ~ 14 dB are obtained with macro-bending at wavelength region between 1480 nm and 1530 nm. This enhancement is attributed to macro-bending effect which suppresses the ASE at the longer wavelength. This resulted in an increase of population inversion at shorter wavelength, which in turn improves the EDFA's gain at the shorter wavelength as shown in Fig. 2. With the macro-bending, the positive gain is observed for input signal wavelength of 1516 nm and above. On the other hand, the macro-bending also reduces the noise figure of the EDFA at wavelengths shorter than 1525nm as shown in Fig. 2.

Fig. 3 shows the gain and noise figure as a function of 980nm pump power with and without the macro-bending. In this experiment, the input signal power and wavelength is fixed at -30 and 1516nm, respectively. The bending radius is fixed at 4 mm. As shown in the figure, the macro-bending improves both gain and noise figure by approximately 6 dB and 3 dB, respectively. These improvements are due to the longer wavelength ASE suppression by the macro-bending effect in the EDF. With the macro-bending, the double-pass EDFA is able to achieve a positive gain with pump power of 90 mW and above. These results show that the bending effect can be used to increase the gain at a shorter wavelength, which has potential applications in S-band EDFA and fiber lasers. The operating wavelength of EDF fiber laser is expected can be tuned to a shorter wavelength region by the macro-bending.

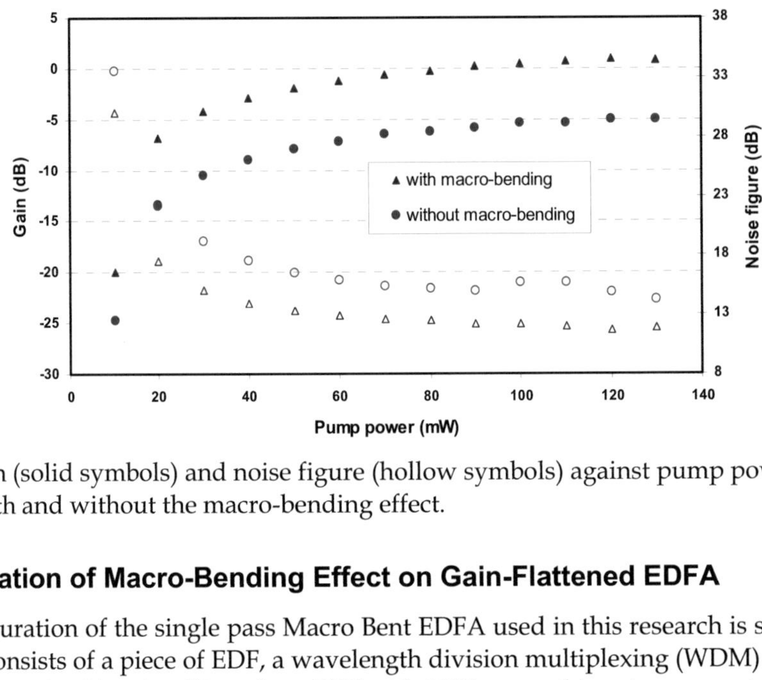

Fig. 3. Gain (solid symbols) and noise figure (hollow symbols) against pump power for EDFAs with and without the macro-bending effect.

3. Application of Macro-Bending Effect on Gain-Flattened EDFA

The configuration of the single pass Macro Bent EDFA used in this research is shown in Fig. 4, which consists of a piece of EDF, a wavelength division multiplexing (WDM) coupler, and a pump laser. An Aluminosilicate host EDF with 1100 ppm erbium ion concentration is used in the setup. Alumina in this fiber is to overcome the quenching effect for high ion concentration. A WDM coupler is used to combine the pump and input signal. Optical isolators are used to ensure unidirectional operation of the optical amplifier. Laser pump power at 980nm is used for providing sufficient pumping power. The EDF is spooled on a rod of 6.5 mm radius to achieve consistent macro-bending effect. The rod has equally spaced threads (8 threads per cm) where each thread houses one turn of EDF to achieve consistency in the desired bending radius. Tunable laser source (TLS) is used to characterize the amplifier in conjunction with the optical spectrum analyzer (OSA) (Hajireza, et al. 2010).

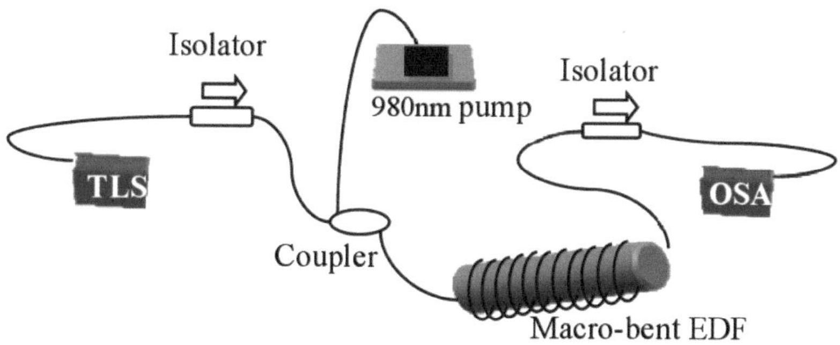

Fig. 4. Configuration of the single-pass EDFA

Initially, the gain and noise figure of the single pass EDFA is characterized without any macro-bending at different EDF lengths. The input signal power is fixed at -30 dBm and the 980 nm pump power is fixed at 200 mW. The wavelength range is chosen between 1520 nm and 1570 nm which covering the entire C-band. It is important to note that using macro-bending to achieve gain flatness depend on suppression of longer wavelength gains. The EDF length used must be slightly longer than the conventional C-band EDFA to allow an energy transfer from C-band to L-band taking place. This will reduce the gain peak at 1530nm and increases the gain at longer wavelengths. The macro-bending provides a higher loss at the longer wavelengths and thus flattening the gain spectrum of the proposed C-band EDFA. The combination of appropriate EDF length and bending radius, leads to flat and broad gain profile across the C-band region.

The bending loss spectrum of the EDF is measured across the wavelength region from 1530 nm to 1570 nm. Fig. 2 illustrates the bending loss profile at bending radius of 4.5 mm, 5.5 mm and 6.5 mm, which clearly show an exponential relationship between the bending loss and wavelength, with strong dependencies on the fiber bending radius. Bending the EDF causes the guided modes to partially couple into the cladding layer, which in turn results in losses as earlier reported. The bending loss has a strong spectral variation because of the proportional changes of the mode field diameter with signal wavelength (Giles et al., 1991) As shown in Fig. 5, the bending loss dramatically increases at wavelengths above 1550 nm. This result shows that the distributed ASE filtering can be achieved by macro bending the EDF at an optimally chosen radius. This provides high ASE or gain suppression around 1560 nm, which reduces the L-band gain. Besides this, lower level suppression of C-band population inversion reduces the effect of gain saturation, providing better C-band gain. Eventually, this characteristic is used to achieve C-band gain flattening in the EDFA (Hajireza, et al. 2010).

The gain spectrum of the EDFA is then investigated when a short length of high concentration EDF spooled in different radius. Fig. 6 shows the gain spectrum of the EDFA with 3m long EDF at different spooling radius. The result was also compared with straight EDF. The input signal power and pump power are fixed at -30dBm and 200 mW respectively in the experiment. As shown in the figure, the original shape of the gain spectrum is maintained in the whole C-band region with the gain decreases exponentially at wavelengths higher than 1560nm. Without bending, the peak gain of 28dB is obtained at 1530 nm which is the reference point to find the optimized length. When the EDF was spooled at a rod with 4.5mm and 5.5mm radius, the shape of gain spectra are totally

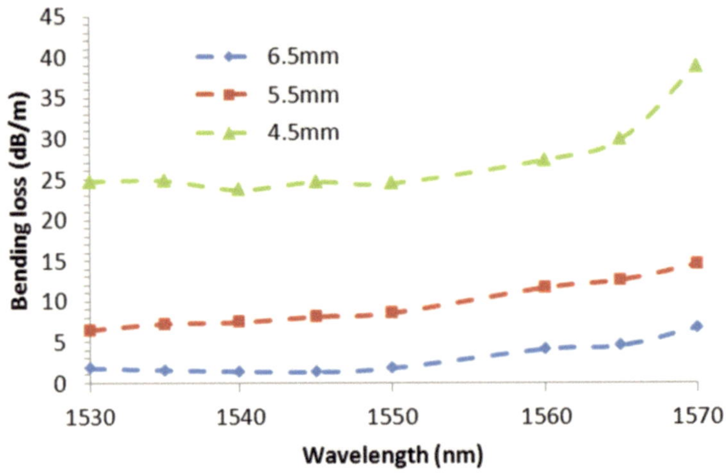

Fig. 5. Bending loss (dB/m) vs. wavelength (nm) for different bending radius

changed. Finally after trying different radius, 6.5 mm was the optimized radius for this amplifier. As shown in Fig. 5, bending loss at radius of 6.5 mm is low especially at wavelengths shorter than 1560nm and therefore the gain spectrum maintains the original shape of the gain spectrum for un-spooled EDF (Hajireza, et al. 2010)..

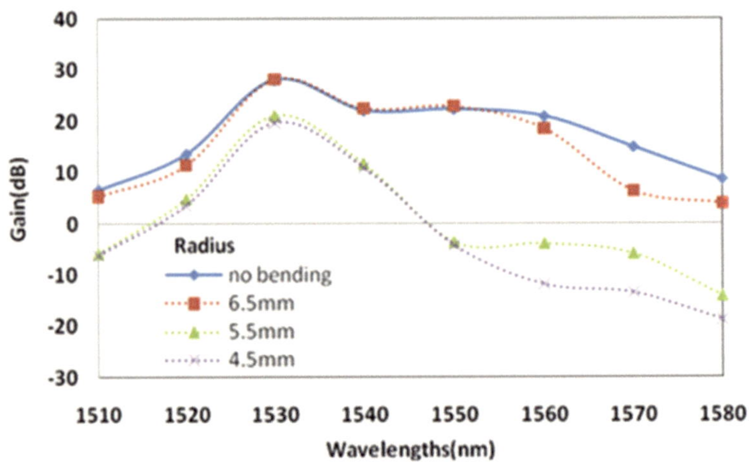

Fig. 6. Efficient length of EDF (3m) in different bending radius for -30dBm input signal]

Fig. 7 shows the gain spectra for the single pass EDFA with unspooled EDF at various EDF lengths. The input signal and 980nm pump powers are fixed at -30 dBm and 200 mW. As the length of the EDF increases, the gain spectrum moves to a longer wavelength region. The C-band photons are absorbed to emit photons at longer wavelength. The overall gain drops at the maximum length of 11m due to the insufficient pump power. Fig. 5 shows the gain spectra for the EDFA with the optimum spooling radius of 6.5 at various EDF lengths.

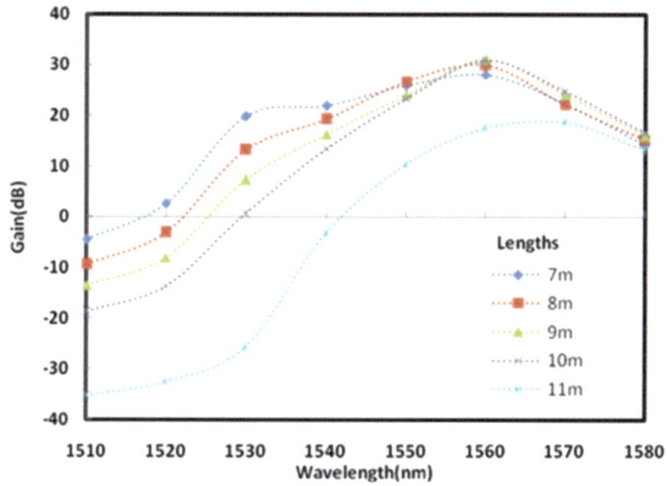

Fig. 7. Gain for un-spooled EDF in different length with -30dBm input signal

To achieve a flatten gain spectrum, the EDFA must operate with insufficient 980nm pump, where the shorter wavelength ASE is absorbed by the un-pumped EDF to emit at the longer wavelength. This will shift the peak gain wavelength from 1530nm to around 1560nm. The macro-bending induces a wavelength dependent bending loss that results in higher loss at longer wavelength compared to the shorter wavelength as shown in Fig. 5. In relation to the EDFA, the macro-bending also suppresses the population inversion in C-band and thus reduces the gain saturation effect in C-band. With this reduced gain saturation, the C-band gain will increase. On the other hand, the L-band gain will reduce due to the suppression of L-band stimulated emission induced by macro-bending. The net effect of both phenomena will result in a flattened gain profile as shown in figure 8. Thus, the level of population inversion is dependent on different parameters such as length of fiber, bending radius and erbium ion concentration of the EDF. The same mechanism of distributed ASE filtering is used for S-band EDFA (Wysocki et al., 1998).

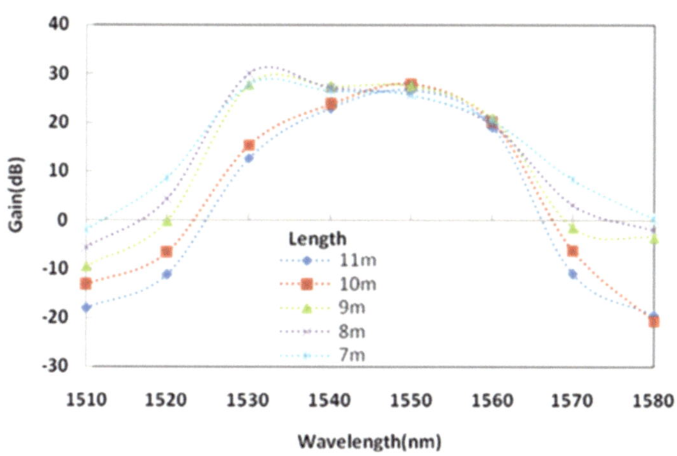

Fig. 8. Gain for 6.5mm spooled radius in different length with -30dBm input signal

Fig. 8 shows a better flattening approach for 9m EDF, where the flattened gain profile is obtained by incremental gain enhancement of about 3 dB and 20 dB at wavelengths of 1550 nm and 1530 nm respectively. This enhancement is attributed to macro-bending effect in the EDF. This incremental gain compensates the incremental gain reduction of the EDFA before applying macro-bending resulting in a flat and broad gain spectrum.
The gain variation to gain ratio ΔG/G is generally used to characterize the gain variation, where ΔG and G are the gain excursion and the average gain value, respectively (Wysocki et al., 1997). In order to define the gain flatness of EDFAs, the ΔG/G for the EDFA with and without macro bending is compared between 1530 and 1555 nm under the same condition. The gain variation ΔG/G for this macro-bent EDFA was 0.10 (2.8 dB / 27.84 dB), which is a 50% improvement compared to earlier reports (Uh-Chan et al., 2002). Besides this, we also observe a gain variation within ±1 dB over 25 nm bandwidth in C-band region (S.D.Emami et al., 2009).
Fig. 9 compares the gain spectrum of the EDFA with and without macro bending EDF at various input signal power. The input signal power is varied for -10 dBm and -30 dBm. The input pump power is fixed at 200 mW. The EDF length and bending radius is fixed at 9m and 6.5 mm respectively. As shown in the figure, increasing the input signal power decreases the gain but improves the gain flatness. The macro bending also reduces the noise figure of EDFA at wavelength shorter than 1550 as shown in Fig. 10 (Hajireza, et al. 2010).

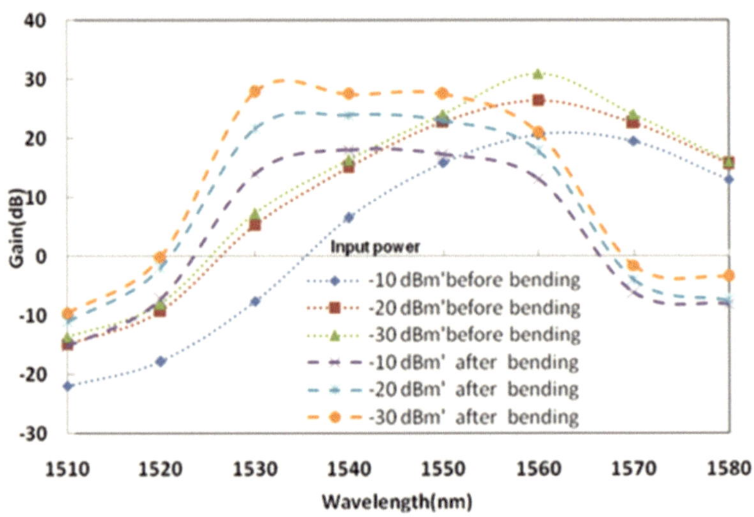

Fig. 9. EDFA gain spectra; with and without acro bending at various input signal power.

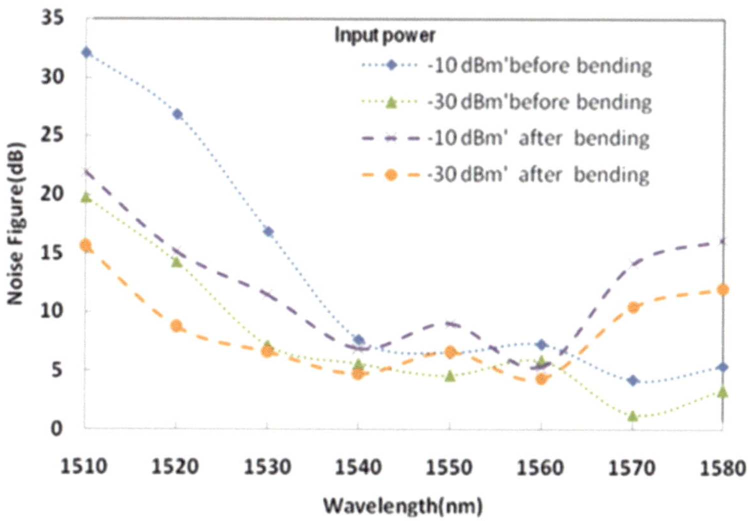

Fig. 10. EDFS noise figure spectra; with and without macro bending at various input signal power.

4. Modelling of the macro-bent EDFA

Macro-bending is defined as a smooth bend of fiber with a bending radius much larger than the fiber radius (Marcuse, 1982). Macro-bending modifies the field distribution in optical fibers and thus changes the spectrum of the wavelength dependent loss. Various mathematical models have been suggested to calculate the bending effects in optical waveguide. Earlier references for bending loss in single mode fibers with step index profiles was developed by Marcuse. According to Marcuse, the total loss of a macro bent fiber includes the pure bending loss and transition loss caused by mismatch between the quasi-mode of the bending fiber and the fundamental mode of the straight fiber (Marcuse, 1976). The analytical expression for a single meter fiber bend loss α can be expressed as follows (Marcuse, 1982):

$$\alpha(\nu) = \frac{\sqrt{\pi}k^2 \exp\left[-\frac{2}{3}\left(\gamma^3 / \beta_g^2\right)R\right]}{e_\nu \gamma^{2/3} V^2 \sqrt{R} K_{\nu-1}(\gamma a) K_{\nu+1}(\gamma a)} \tag{1}$$

where ev=2 , a is the radius of fiber core, R is the bending radius, βg is the propagation constant of the fundamental mode, K(υ-1)(γα) and K(υ+1)(γα) are the modified Bessel functions and V is the well-known normalized frequency, which is defined as (Agrawal, 1997):

$$V = \frac{2\pi a.NA}{\lambda} \tag{2}$$

The values of k and γ can be defined as follows (Gred & Keiser, 2000):

$$k = \sqrt{n_1^2 k^2 - \beta_g^2} \tag{3}$$

$$\gamma = \sqrt{\beta_g^2 - n_2^2 k^2} \tag{4}$$

For an optical fiber with length L, bending loss (α) is obtained by: $\alpha_L = 10\log(\exp(2\alpha l)) = 8.68 aL$ Equation (3) agrees well with our experimental results for macro-bent single-mode fiber.

The macro-bent EDFA is modeled by considering the rate equations of a three level energy system. Fig.11 shows the absorption and emission transitions, respectively in the EDFA considering a three-level energy system with 980 nm pump. Level 1 is the ground level, level 2 is the metastable level characterized by a long lifetime, and level 3 the pump level (Armitage, 1988) . The main transition used for amplification is from the 4I13/2 to 4I15/2 energy levels. When the EDF is pumped with 980 nm laser, the ground state ions in the 4I15/2 energy level can be excited to the 4I11/2 energy level and then relaxed to the 4I13/2 energy level by non-radiative decay. The variables N_1, N_2 and N_3 are used to represent population of ions in the ${}^4I_{15/2}$, ${}^4I_{13/2}$ and ${}^4I_{11/2}$ energy levels respectively. According to Fig. 11 we can write the rate of population as follows (Desurvire, 1994):

$$\frac{dN_1}{dz} = -R_{13}N_1 + R_{31}N_3 - W_{12}N_1 + W_{21}N_2 + {}^{R}A_{21}N_2 \tag{5}$$

$$\frac{dN_2}{dt} = -W_{12}N_1 - W_{21}N_2 - {}^{R}A_{21}N_2 + {}^{NR}A_{23}N_3 \tag{6}$$

$$\frac{dN_3}{dz} = R_{13}N_1 - R_{31}N_3 - {}^{NR}A_{32}N_3 \tag{7}$$

$$N_T = N_1 + N_2 + N_3 \tag{8}$$

where R_{13} is the pumping rate from level 1 to level 3 and R_{31} is the stimulated emission rate between level 3 and level 1. The radiative and non radiative decay from level i to s is represented by ${}^{R}A_{ij}$ and ${}^{NR}A_{ij}$. The interaction of the electromagnetic field with the ions or the stimulated absorption and emission rate between level 1 and level 2 is represented by W_{12} and W_{21}.

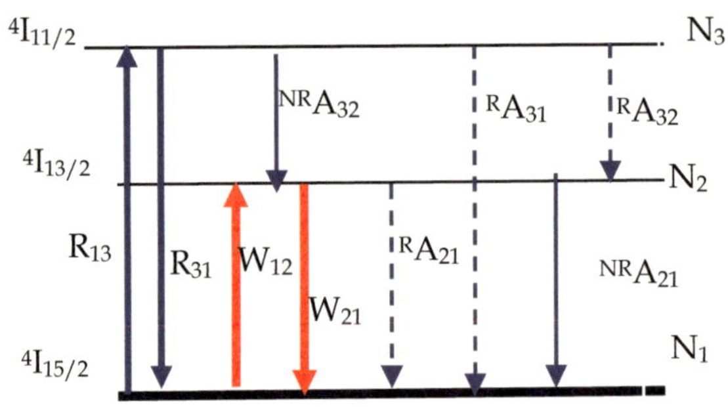

Fig. 11. Three level energy system of EDF pump absorption, and signal transitions.

The stimulated absorption, emission rate and pumping rate are calculated respectively as follows (Desurvire et al., 1990) :

$$W_{12} = \frac{\sigma_{SA}(\lambda_s)\Gamma_s}{h\nu_s A}\left[P_S + P_{ASE}^{+} + P_{ASE}^{-}\right] \tag{9}$$

$$W_{21} = \frac{\sigma_{SE}(\lambda_s)\Gamma_s}{h\nu_s A}\left[P_S + P_{ASE}^{+} + P_{ASE}^{-}\right] \tag{10}$$

$$R = \frac{\sigma_{PA}(\lambda_P)\Gamma_P}{h\nu_p A}\left[P_P\right] \tag{11}$$

where σ_{PA} is the $^4I_{11/2} \rightarrow {}^4I_{15/2}$ absorption cross sections of the 980 nm forward pumping. σ_{SA} and σ_{SE} are the stimulated absorption and the stimulated emission cross-section of input signal respectively. PASE is the amplified spontaneous emission (ASE) power and A is the effective area of the EDF. The light-wave propagation equations along the erbium-doped fiber can be established as follows (Parekhan et al., 1988):

$$\frac{dP_{ASE}^{\pm}}{dz} = \pm\Gamma(\lambda_{ASE})(\sigma_{se}N_2 - \sigma_{se}N_1)\times P_{ASE}^{\pm} \pm \Gamma(\lambda_s)2h\nu\Delta\nu\sigma_{se}N_2 \mp \alpha P_{ASE}^{\pm} \tag{12}$$

$$\frac{dP_P^{+}}{dz} = \Gamma(\lambda_P)(\sigma_{PE}N_2 - \sigma_{PA}N_1)\times P_P^{+} - \alpha P_P^{+} \tag{13}$$

$$\frac{dP_s^{+}}{dz} = \Gamma(\lambda_s)(\sigma_{se}N_2 - \sigma_{sa}N_1)\times P_s - \alpha P_s \tag{14}$$

Absorption and emission coefficient are essential parameters to know for any types of EDFA modelling. With aid of cutback method the absorption coefficient of fiber was measured experimentally (Hajireza, et al. 2010). For an EDF with uniform radial core doping it is preferred to use the MFD expression developed by (Myslinski et al.,1996). The absorption cross section and emission cross section in room temperature were calculated respectively as follows:

$$\alpha(\lambda) = \sigma a(\lambda)\Gamma(\lambda)\mathrm{nt} \tag{15}$$

$$\sigma a(\upsilon) = \sigma e(\upsilon)\exp\left(\frac{h\nu - E_0}{K_B T}\right) \tag{16}$$

$\sigma_a(\lambda)$ is absorption Cross section that describes the chance of an erbium ion absorbing a photon at wavelength λ. Cross section is given in terms of area because it represents the area is occupied by each erbium ion ready to absorb. Multiplying this by the number of ions, s, gives the total area of the fiber cross section that has erbium ready to absorb. The overlapping factors between each radiation and the fiber fundamental mode, Γ (λ) can be expressed as (Desurvire, 1990):

$$\Gamma(\lambda) = 1 - e^{-\frac{2b^2}{\omega_0^2}} \tag{17}$$

$$\omega_0 = a\left(0.761 + \frac{1.237}{V^{1.5}} + \frac{1.429}{V^6}\right) \tag{18}$$

where ω_0 is the mode field radius defined by equation (18), a is the core diameter, b is the Erbium ion-dopant radius and V is the normalized frequency. The absorption and emission cross section has shown in fig.12 (Michael & Digonnet, 1990). Background scattering loss and wavelength-dependent bending loss is represented by α (λ). Wavelength-dependent bending losses used in this numerical model for three different bending diameters as shown in Fig. 13. The bending loss spectral profile is obtained theoretically with help of Marcuse formula (Marcuse, 1982).. These bending radius values are chosen because significant bending losses can be observed in the L-band region. The bending loss profile indicates the total distributed loss for different bending radius associated with macro-bending at different EDF lengths. This information is important when choosing the appropriate bending radius to achieve sufficient suppression of the gain saturation effect in L-band region and reduces the energy transfer from C-band to the longer wavelength region (Giles & Digiovanni, 1990).

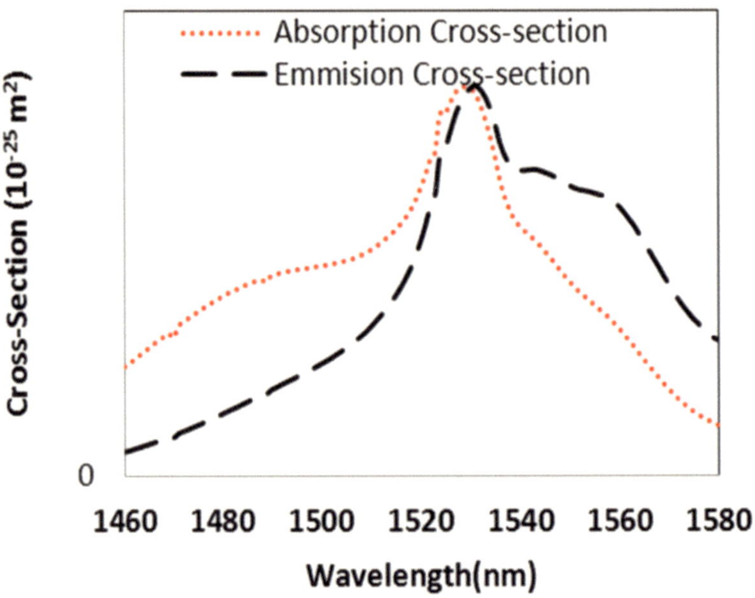

Fig. 12. Absorption and Emission Cross section.

In order to solve the population rate in steady state condition, the time derivatives of for pump and signal powers, equations are set to zero. All the equations are first order differential equations and the Runge-Kutta method is used to solve these equations. The variables used in the numerical calculation and their corresponding values are shown in Table 1.

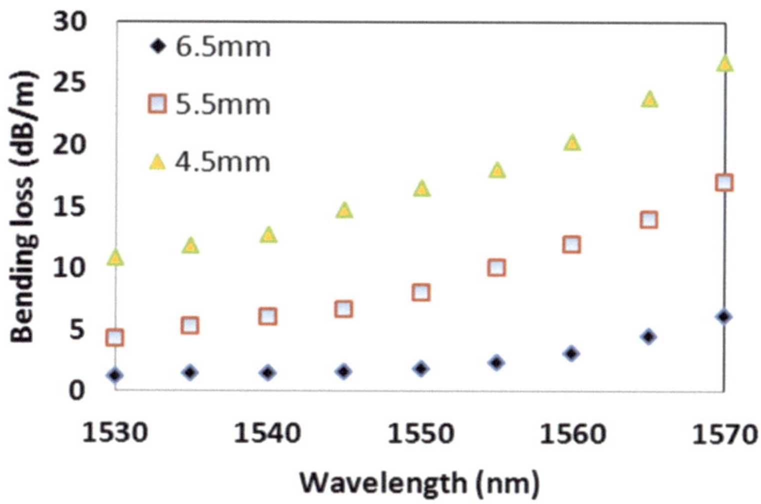

Fig. 13. Bending loss spectral for different bending radius

Parameter	Unit	Value
NA (typ)	0.22	-
λcut-off (typ)	935	[nm]
dcore (typ)	3.3	[μm]
Doping density	1.6	[×1025 ions/m3]
τ (Life time)	10	[ms]
Saturation paramter (typ)	7.985	[×1015 /ms]
λpump	980	[nm]
MFDpump	3.7	[μm]
A	$1.633\text{x}10^{-11}$	m^2
nclad	1.451	-
ncore	1.469	-
λsig	1550	[nm]
MFDsig	5.3	[μm]
Γsig	0.74	-
Γpump	0.77	-
$\sigma_{SA}(\lambda_s)$	$2.9105\text{x}10^{-25}$	m^2
$\sigma_{SE}(\lambda_s)$	$4.1188\text{x}10^{-25}$	m^2
$\sigma_{PA}(\lambda_p)$	$2.78\text{x}10^{-25}$	m^2
$\sigma_{PE}(\lambda_p)$	$0.81056\text{x}10^{-25}$	m^2
Δv	3100	GHz

Table 1. Numerical parameter used in the simulation

The bending loss spectrum of the EDF is measured across the wavelength region from 1530 nm to 1570 nm. Fig. 14 illustrates the bending loss profile at bending radius of 4.5 mm, 5.5 mm and 6.5 mm, which clearly show an exponential relationship between the bending loss and wavelength, with strong dependencies on the fiber bending radius. Bending the EDF causes the guided modes to partially couple into the cladding layer, which in turn results in losses as earlier reported. The bending loss has a strong spectral variation because of the proportional changes of the mode field diameter with signal wavelength. As shown in Fig. 14, the bending loss dramatically increases at wavelengths above 1550 nm. This result shows that the distributed ASE filtering can be achieved by macro bending the EDF at an optimally chosen radius. It was important to analysis the bending loss in an optimized C-band amplifier before proceed to the next step. The results as shown in Fig. 15 indicate that 3 meter is optimized length for C- band amplifier. It was also seen that with decreasing length, S-band gain is increasing. This happened because of reduction of inversion in C-band region which allow a peak competition for S-band photons to increase. In general C-band always keeps the gain peak unless for longer lengths.

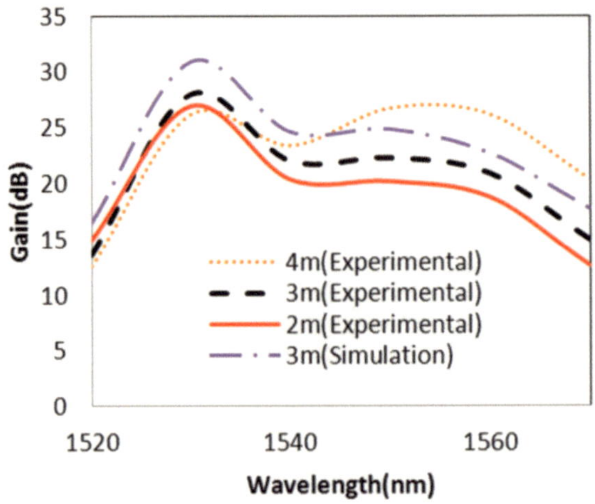

Fig. 14. Different length of unspooled EDFA for -30dBm input signal

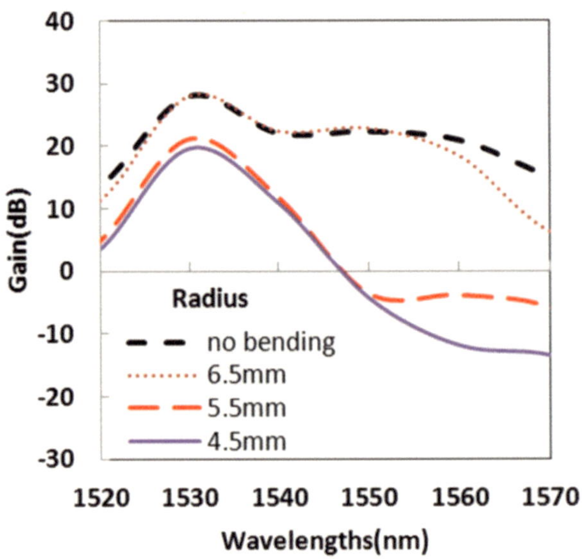

Fig. 15. Efficient length of EDF (3m) in different bending radius for -30dBm input signal (Experimental).

The gain spectrum of the EDFA is then investigated when the optimized length of high concentration EDF spooled in different radius. Fig. 16 shows the gain spectrum of the EDFA with 3m long EDF at different spooling radius. The result was also compared with straight EDF. The input signal power and pump power are fixed at -30dBm and 200 mW respectively in the experiment. As shown in the figure, the original shape of the gain spectrum is maintained in the whole C-band region with the gain decreases exponentially at wavelengths higher than 1560nm. Without bending, the peak gain of 28dB is obtained at 1530 nm which is the reference point to find the optimized length. When the EDF was spooled at a rod with 4.5mm and 5.5mm radius, the shape of gain spectra are totally changed. Finally after trying different radius, 6.5 mm was the optimized radius for this amplifier.

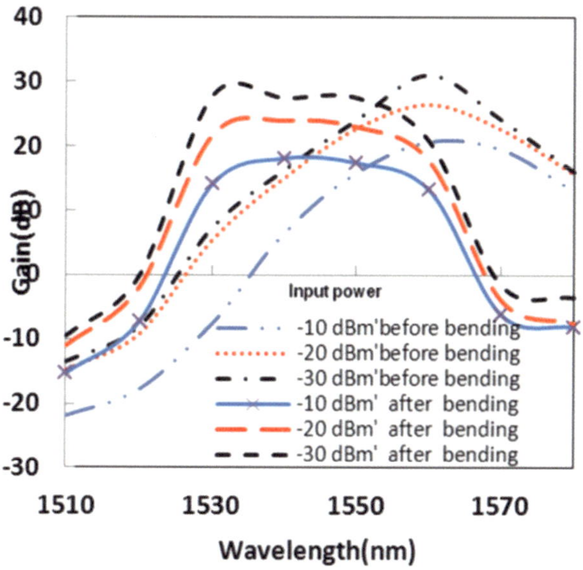

Fig. 16. Gain profile of EDFA with and without macro bending at various input signal power.(Experimental)

As shown in Fig. 14, bending loss at radius of 6.5 mm is low especially at wavelengths shorter than 1560nm and therefore the gain spectrum maintains the original shape of the gain spectrum for un-spooled EDF. To achieve a flatten gain spectrum, the unbent EDFA must operate with insufficient 980nm pump, where the shorter wavelength ASE is absorbed by the un-pumped EDF to emit at the longer wavelength. This will shift the peak gain wavelength from 1530nm to around 1560nm. The macro-bending induces bending loss is dependent on wavelength with an exponential relationship and longer wavelength has a higher loss compared to the shorter wavelength. In relation to the EDFA, the macro-bending also increase the population inversion in C-band due to reduction of gain saturation effect in L-band. Since the L-band gain cannot improve more than a limited value due to exposure bending loss, less C band photons will be absorbed by un-pumped ions to emit at L-band. This effect reduced gain saturation in L-band, so the C-band gain will increase. This increment for peak is not more than the optimized C-band EDFA (3m) since at that level the inversion is in the maximum value. Full inversion for bent EDFA take place at longer length due to limited energy transfer to longer wavelength. On the other hand, the L-band gain will reduce due to the suppression of L-band stimulated emission induced by macro-bending. The net effect of both phenomena will result in a flattened gain profile.

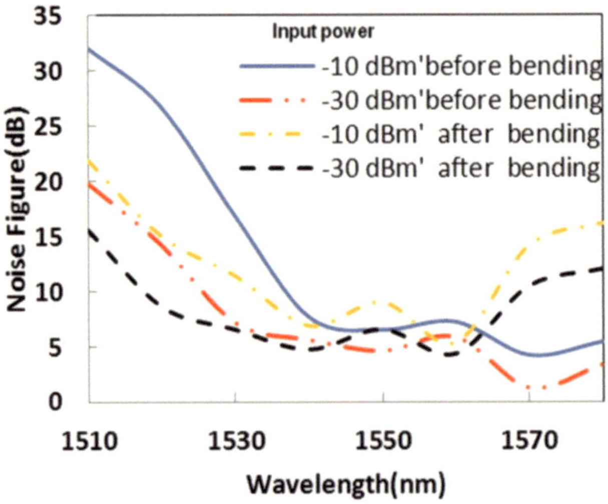

Fig. 17. Noise figure profile of EDFA with and with-out macro bending at various input signal power(Experimental)

Fig. 17 compares the gain spectrum of with and with-out macro bending EDF at various input signal power. The input signal power is varied for -10 dBm to -30 dBm. The input pump power is fixed at 200 mW. The EDF length and bending radius is fixed at 9 meter and 6.5 mm respectively. As shown in the figure, increasing the input signal power decreases the gain but improves the gain flatness. The macro bending also reduces the noise figure of EDFA at wavelength shorter than 1550 as shown in Fig. 18. Since keeping the amount of noise low depends on a high population inversion in the input end of the erbium-doped fiber (EDF), the backward ASE power P -ASE is reduced by the bending loss. Consecutively, the forward ASE power P +ASE can be reduced when the pump power P is large at this part of the EDF which is especially undesirable. This is attributed can be described numerically by the following equation (Harun et al., 2010)

$$NF = \frac{1}{G} + \frac{2P_{ASE}}{Gh\nu} \tag{19}$$

where G is the amplifier's gain, PASE is the ASE power and hv is the photon energy.
Fig 19 indicates the simulation of ASE for standard C-band EDFA (3m) and optimized gain flattened C-band EDFA (9m) after and before bending. ASE here represents population inversion. It clearly explains the gain shifting from longer wavelength to the shorter wavelength due to length increment. Besides effect of bending on gain flattening is explained. Fig 10 is the comparison between standard C-band EDFA with the flattened gain EDFA. We observe a gain variation within ±1 dB over 25 nm bandwidth in C-band region.

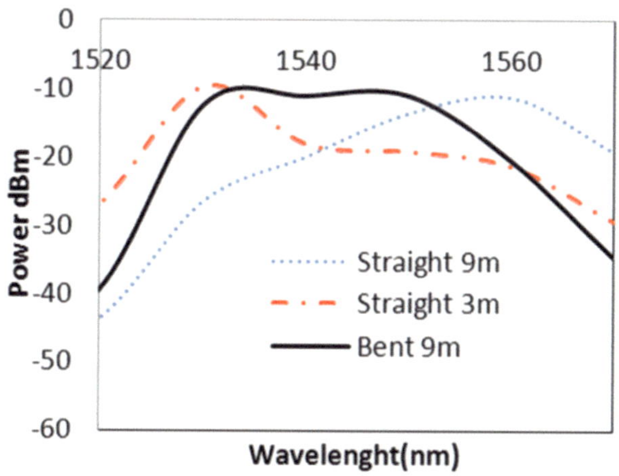

Fig. 18. Amplified Spontaneous emission (Simulation)

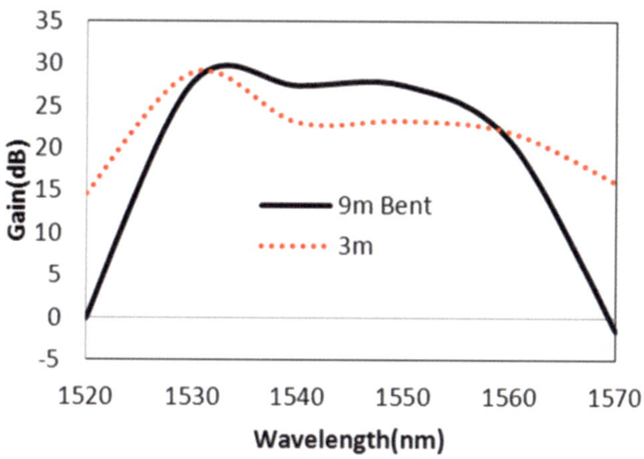

Fig. 19. Comparison of the standard C-band EDFA with the flattened gain EDFA for -30 dB input power(Experimental)

5. Temperature insensitive broad and flat gain EDFA based on macro-bending

Recently, macro-bent EDF is used to achieve amplification in S-band region. In this paper, a gain-flattened C-band EDFA is proposed using a macro-bent EDF. This technique is able to compensate the EDFA gain spectrum to achieve a flat and broad gain characteristic based on distributed filtering using a simple and low cost method. This technique is also capable to compensate the fluctuation in operating temperatures due to proportional temperature sensitivity of absorption cross section and bending loss of the aluminosilicate EDF. This new approach can be used to design a temperature insensitive EDFA for application in a real optical communication (Hajireza, et al. 2010).

The bending loss profile of the erbium-doped fiber (EDF) for various bending radius is firstly investigated by conducting a simple loss- test measurement. In order to isolate the bending loss, the profile is obtained by taking the difference between the loss profile of the same EDF with and without macro-bending across the desired wavelength range. A one meter EDF is used in conjunction with a tunable laser source (TLS) and optical power meter to characterize the bending loss for bending radius of 6.5 mm at wavelength region between 1530 nm and 1570 nm. The bending loss profile indicates the total distributed loss for different bending radius associated with macro-bending at different EDF lengths. This information is important when choosing the optimized bending radius to achieve sufficient suppression of the gain. Fig. 20 illustrates the bending loss profile at bending radius of 6.5 mm at different temperatures, which clearly show an exponential relationship between the bending loss and wavelength. It is also shown that the bending loss in L-band is reduced by increasing the temperature. Bending the EDF causes the guided modes to partially couple into the cladding layer, which in turn results in losses as earlier reported. The bending loss has a strong spectral variation because of the proportional changes of the mode field diameter with signal wavelength. As shown in Fig. 20, the bending loss dramatically increases at wavelengths above 1550 nm. This result shows that the distributed ASE filtering can be achieved by macro bending the EDF at an optimally chosen radius.

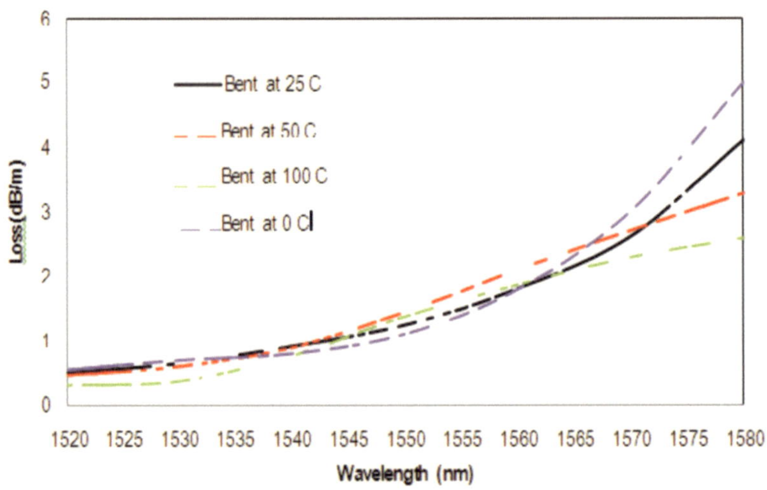

Fig. 20. Loss spectrum of the bent EDF with 6.5 mm bending radius at different temperatures.

Initially, the gain of the single pass EDFA is characterized without any macro-bending at different EDF lengths as shown in Fig. 21. The input signal power is fixed at −30 dBm and the 980 nm pump power is fixed at 200 mW. The wavelength range is chosen between 1520 nm and 1570 nm which cover the entire C-band region. To achieve a flattened gain spectrum, the unbent EDFA must operate with insufficient 980 nm pump, where the shorter wavelength ASE is absorbed by the un- pumped EDF to emit at the longer wavelength. This will shift the peak gain wavelength from 1530 nm to around 1560 nm. Therefore The EDF length used must be slightly longer than the conventional C- band EDFA to allow an energy transfer from C-band to L-band taking place. This will reduce the gain peak at 1530 nm and increases the gain at longer wavelengths. As shown in Fig. 21, the optimum C-band operation is successfully achieved using only one meter of this high erbium ion concentration EDF. It is also shown that for the lengths longer than 2m gain shifts to longer wavelengths. Figure 22 shows the gain spectrum of the C-band EDFA, which is characterized with macro-bending at different EDF lengths. In the experiment, the input signal power is fixed at -30dBm and the 980nm pump power is fixed at 200mW. These lengths are chosen due to their gain shift characteristics as depicted in Fig. 21. It is important to note that using macro-bending to achieve gain flatness depend on suppression of longer wavelength gains. The macro-bending provides a higher loss at the longer wavelengths and thus flattening the gain spectrum of the proposed C-band EDFA. The combination of appropriate EDF length and bending radius, leads to flat and broad (Hajireza, et al. 2010).

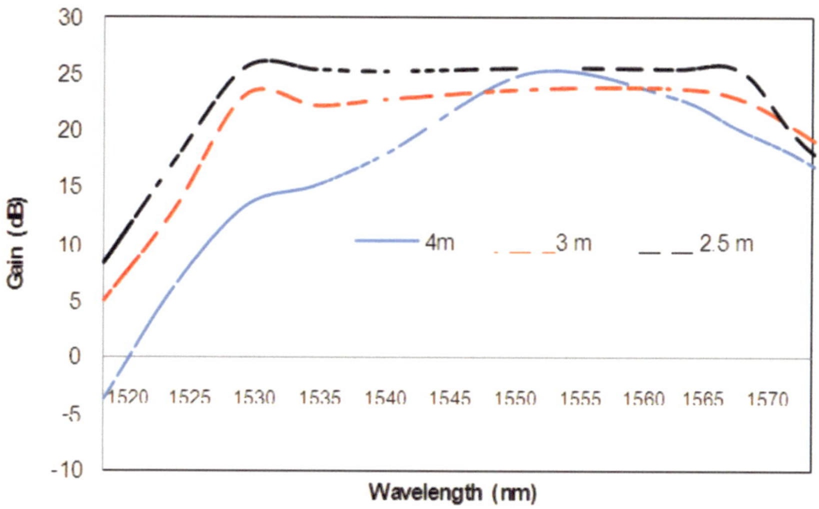

Fig. 21. Gain spectrum of the C-band EDFA

6. Effects of erbium transversal distribution profiles on EDFA performance

Over the past years, Erbium-doped fiber amplifiers (EDFAs) have received great attention due to their characteristics of high gains, bandwidths, low noises and high efficiencies. As a key device, EDFA configures wavelength division multiplexing systems (WDMs) in optical telecommunications, finding a variety of applications in traveling-wave fiber amplifiers, nonlinear optical devices and optical switches. The EDFA uses a fiber whose core is doped with trivalent erbium ions as the gain medium to absorb light at pump wavelengths of 980 nm or 1480 nm and emit at a signal wavelength band around 1500 nm through stimulated emission. Theoretical study on optimization of rare-earth doped fibers, such as fiber length and pump power has grown along with their increased use and greater demand for more efficient amplifiers (Emamai et al., 2010). Previously, one of the most important issues in improving fiber optic amplifier performance is optimization of the rare-earth dopant distribution profile in the core of the fiber. Earlier approaches to numerical modeling of EDFA performance have assumed that the Erbium Transversal Distributions Profile (TDF) follow a step profile.

Only the portion of the optical mode which overlaps with the erbium ion distribution will stimulate absorption or emission from erbium transitions. The overlap factor equation is defined by (Desurvire, 1982):

$$\Gamma(\lambda) = \frac{2\pi}{N_T}\int_0^\infty \Psi(r,v)\times n_T(r)\times r\times dr \tag{20}$$

$\Psi(r, \upsilon)$ is the LP01 fiber optic mode envelope, which is almost Gaussian and is defined as :

$$\Psi(r,v) = \begin{cases} j_0^2(u_k r/a) & r \le a \\ \dfrac{j_0^2(u_k)}{K_0^2(w_k)} K_0^2(w_k r/a) & r \ge a \end{cases} \tag{21}$$

where J0 and K0 are the respective Bessel and modified Bessel functions and uk and wk are the transverse propagation constants of the LP_{01} mode. N_T is total dopant concentration per unit per length which is defined by:

$$N_T = 2\pi\int_0^\infty n_T(r)\times r\times dr \tag{22}$$

Various profiles of erbium transversal distributions can be used for describing mathematical function of EDFA. Two main requirement on choosing erbium transversal distribution functions are; flexibility to be adapted to a collection of profile as broad as possible and dependence on a number of the parameters as low as possible. The optimum transversal distribution function should be (Yun et al., 1999):

$$n_T(r) = n_{T,\max}\exp\{-[|r-\delta|/\theta]^\beta\} \tag{23}$$

where nT,maz is the value of the maximum erbium concentration per unit volume. β, θ and δ are distribution profile parameters which construct the profile shapes. θ and β are defined as dopant radius and the roll-off factor of the profile respectively. In practical, it would seem difficult to maintain a high concentration of Erbium in the center of the core, due to diffusion of erbium ions during the fabrication. Modifying the δ value, low ion concentration at the core center can be achieved. The Erbium distribution profiles of EDFA with different values β, θ and δ are depicted in Fig. 22. Figure 22(a) shows several of β values with fixed values of θ=1 and δ=1.5. Figure 22(b) shows the effect of θ values with fix values of δ=1.5 and β=1 while figure 22(c) shows several values of δ changes with fixed values of θ=1 and β =1.5.

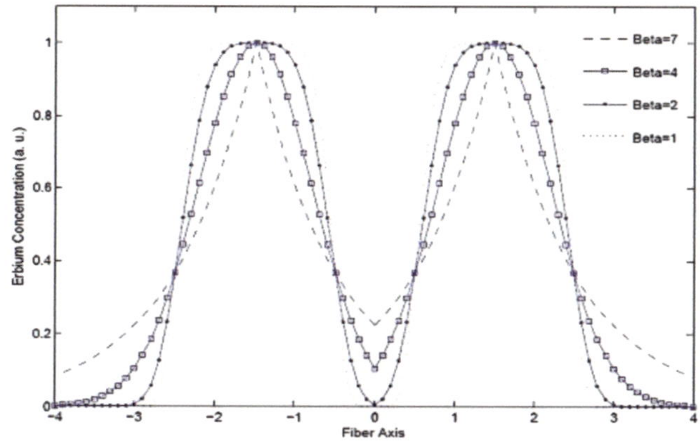

(a)

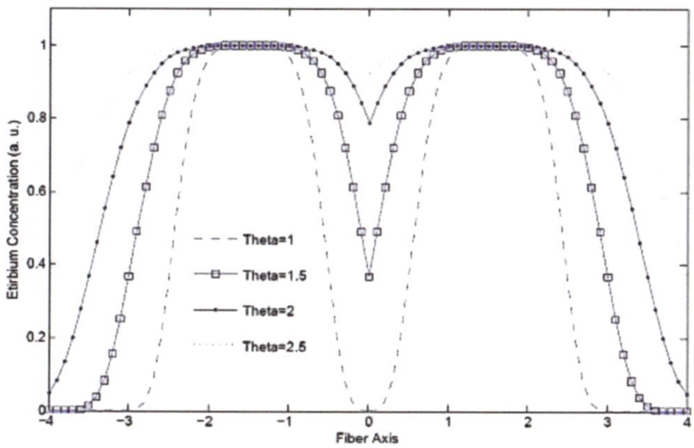

(b)

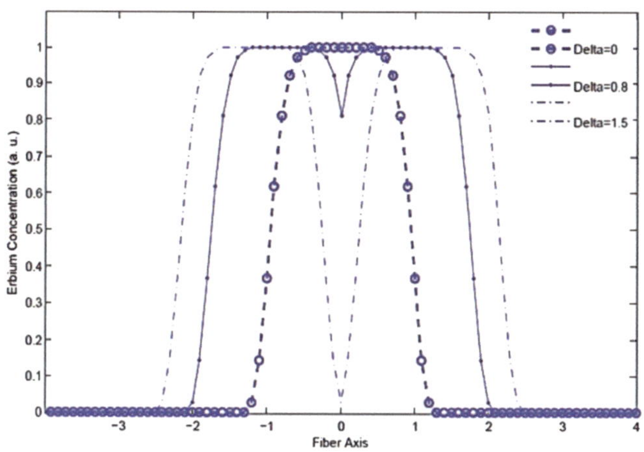

(c)

Fig. 22. The Erbium distribution profile of EDFA (a) at different values of β when, θ=1 and δ=1.5 (b) at different values of θ when δ=1.5 and β=1. (c) at different values of δ when θ=1 and β =1.5.

Figs. 23(a) and (b) demonstrate the gain and noise figure trends of EDFA, respectively at different β and θ values of fiber. The input signal power and input pump power is fixed at -30 dBm and 100 mW respectively while the EDF length is fixed at 14m.

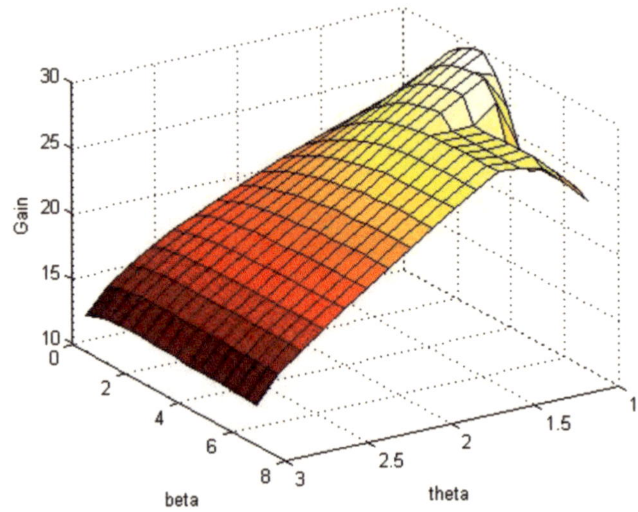

(a)

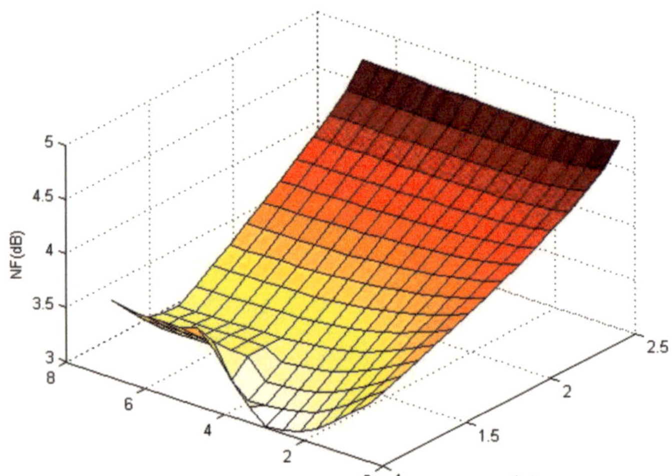

(b)

Fig. 23. (a) Gain profile of EDFA at β=0 (b) Noise figure of EDFA β=0

Fig. 24 shows the gain trends of EDFA at different β and δ values of fiber. The input signal power and input pump power is fixed -30 dBm and 100 mW respectively. The EDF length is 14m long. By comparison between overlap factor and gain results, it is intuitive that the gain result follows the overlap factor values of the fiber. In the low θ values of the fiber in the same EDF concentration the gain decrease by decreasing the θ as depicted on figure 24, this is the results of high erbium intensity at the core and the quenching effect on the fiber amplifier.

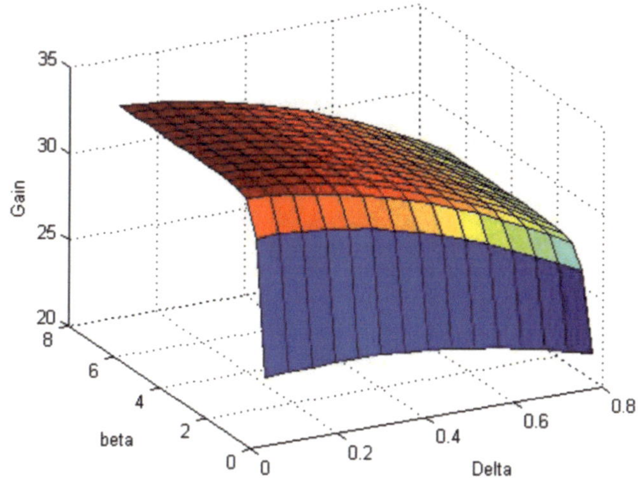

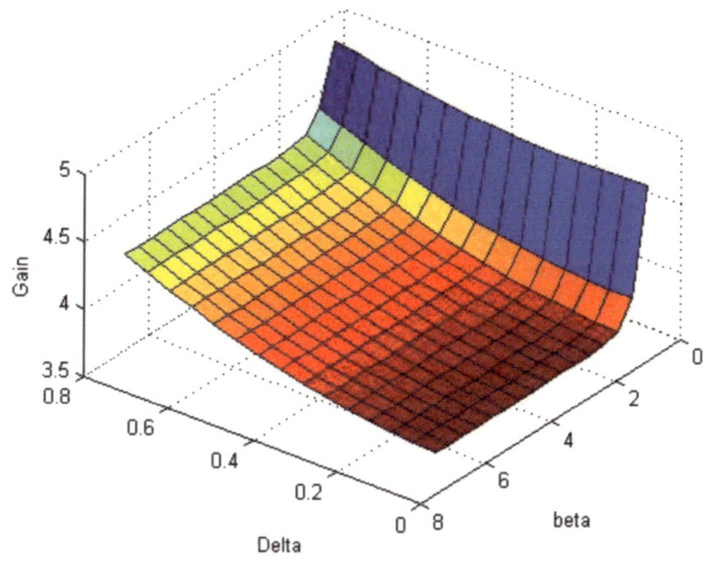

Fig. 24. (a) Gain profile of EDFA at θ=2 (b) Noise figure of EDFA θ=2

The effect of Erbium transversal distribution profile on the performance of an EDFA is investigated. The EDFA uses a 14m long EDF as the gain medium, which is pumped by a 980 nm laser diode via a WDM coupler. An optical isolator is incorporated in both ends of optical amplifier to ensure unidirectional operation. Two types of EDF with the same fiber structure and doping concentration but different on distribution profile are used in the experiment. Fig. 25 shows the Erbium transversal distribution profile of both fibers, which have a doping radius of 2 μm and 4 μm as shown in Figs. 25(a) and (b), respectively. In the experiment, the input signal power and 980nm pump power are fixed at -30dBm and 100 mW respectively.

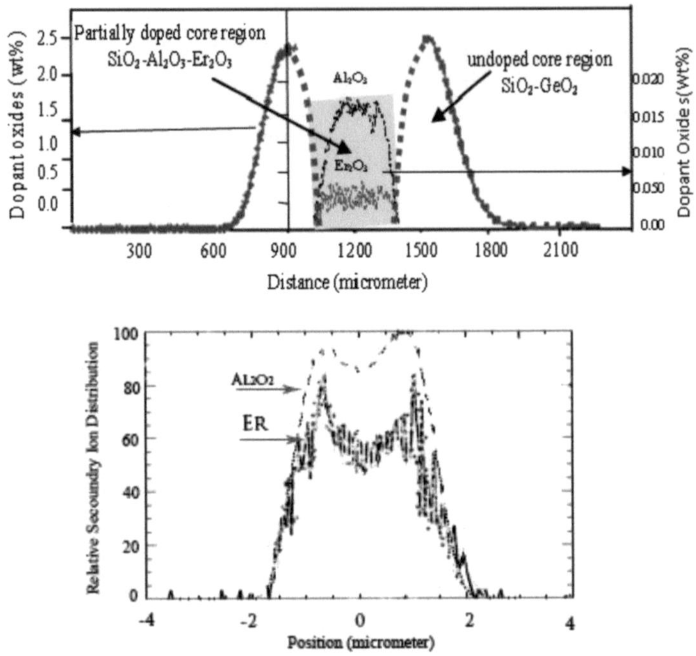

Fig. 25. Erbium TDP. (a) 2μm doping radius (b) 4μm doping

Fig. 26 compares the experimental and numerical results on the gain characteristics for both EDFAs with the 2μm and 4μm doping radius. As expected from the theoretical analysis, the amplifier's gain is higher with 2μm doping radius compared to that of 4μm doping radius at the 1550 nm wavelength region. In the simulation, fiber distribution profile parameters are set as $\theta=2$, $\beta=1.5$ and $\delta=0$ for 2μm doping radius which for 4μm doping radius, fiber distribution profile parameters are set as $\theta=4$, $\beta=4$ and $\delta=0.8$. The numerical gain is observed to be slightly higher than the experimental one. This is most probably due to splicing or additional loss in the cavity, which reduces the attainable gain. In the case of 4 μm dopant radius, the overlap factor is higher since the overlap happens throughout the core region. However, the high overlap factor will affect the erbium absorption of both the pump and signal. If one considers the near Gaussian profile of the LP01 mode, the erbium in the outer radius of the core tend to be less excited due to the lower pump intensity. The remaining Erbium ions absorption capacity in the outer radius of the core will be channeled to absorbing the signal instead. In the case of 2 μm dopant radius, the overlap factor is lower since the overlap happens only in the central part of the core region. If one considers the near Gaussian profile of the LP01 mode, the erbium in the inner radius of the core tend to be more excited due to the higher pump intensity. Since the outer radius of the core is not doped with Erbium, the lower intensity pump in the outer radius will not be absorbed. The advantage of reduced doping region is that the Erbium absorption only takes place in the central part of the core. Since, the pump intensity is the highest here; the Erbium population can be totally inverted, thus contributing to higher gain. Furthermore, the signal in the outer radius will no longer be absorbed. Hence, the signal will receive a net emission from the erbium which then contributes to higher gain.

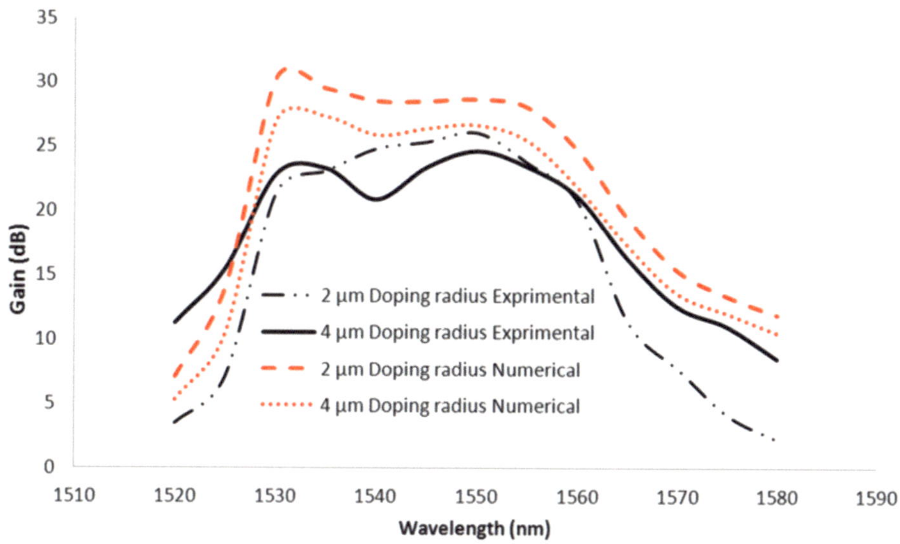

Fig. 26. Numerical and experimental gain comparison of 2μm and 4μm doping radius EDFA

7. Conclusion

In this reserch work a macro-bending approach is demonstrated to increase a gain and noise figure at a shorter wavelength region of EDFA. In the conventional double-pass EDFA configuration , macro-bending improves both gain and noise figure by approximately 6 dB and 3 dB, respectively. These improvements are due to the longer wavelength ASE suppression by the macro-bending effect in the EDF. A new approach is proposed at secound section to achieve flat gain in C-band EDFA with the assistance of macro-bending. The gain flatness is optimum when the bending radius and fiber length are 6.5 mm and 9 meter respectively. This simple approach is able to achieve ±1 dB gain flatness over 25 nm. This cost effective method, which improves the gain variation to gain ratio to 0.1, does not require any additional optical components to flatten the gain, thus reducing the system complexity. The proposed design achieves temperature insensitivity over a range of temperature variation. The gain flatness is optimized when the bending radius and fiber length are 6.5mm and 2.5m respectively. This simple approach is able to achieve 0.5 dB gain flatness over 35nm with no dependency on temperature variations. It is a cost effective method which needs 100mW pump power and does not require any additional optical components to flatten the gain, thus reducing the system complexity. At the end the effect of ETP on the performance of the EDFA is theoretically and experimentally investigated. The ETP can be used to optimize the overlap factor, which affects the absorption and emission dynamics of the EDFA and thus improves the gain and noise figure characteristics of the amplifier. It is experimentally observed that the 1550 nm gain is improved by 3 dB as the doping radius is reduced from 4μm to 2μm. This is attributed to the Erbium absorption takes place in the central part of the core where the pump intensity is the highest and thus increases the population inversion.

8. References

C.R.Giles, C. A. Burrus, D.Digiovanni, N.K. Dutta and G.Raybon. "Characterization of erbium -doped fibers and application to modeling 980nm and 1480nm pumped amplifiers."IEEE photonics technol. Letter, vol 3, no 4, 363, 1991.

C.R.Giles, D.Digiovanni "spectral dependence of gain and noise in erbium doped fiber amplifier" IEEE photonics technol. Lett., vol. 2, no. 11, 1990

D. Marcuse, "Curvature loss formula for optical fibers," Journal Optical Society America B, vol. 66, pp. 216–220, Mar. 1976.

D. Marcuse, "Influence of curvature on the losses of doubly clad fibers," Applied Optics, vol. 21, pp. 4208–4213, 1982.

D. Marcuse, Light Transmission Optics, 2nd. New York: Van Nostrand Reinhold, , pp. 406–414. ,1982.

E. Desurvire, "Erbium-doped fiber amplifiers: principles and applications", John Wiley & Sons, New York, 1994.

E. Desurvire, J. L. Zyskind, C. R. Giles, "Design Optimization for Efficient Erbium-Doped Fiber Amplifiers" Journal of Lightwave Technology, vol. 8. no. 11, November 1990.

G.P. Agrawal, Fiber-Optic Communication Systems, 2nd ed. Wiley, New York, NY, 1997.

Gred, Keiser. Optical Fiber Communications. Singapore: McGraw-Hill, 2000.

Hassani, E. Arzi, F . Seraji "Intensity based erbium distribution for erbium doped fiber amplifiers" Optic Quantum Electron, vol. 39, no. 1, pp: 35-50, 2007.

J. C. Martin"Erbium transversal distribution influence on the effectiveness of a doped fiber: optimization of its performance on the effectiveness of a doped fiber: optimization of its performance" vol. 194, pp. 331-339, 2001

J. Michael, F. Digonnet, "Rare-earth-doped Fiber Lasers and Amplifiers" CRC Press,

J. R. Armitage "Three-level fiber laser amplifier: a theoretical model", APPLIED OPTICS, Vol. 27, no. 23, 1988

K. Thyagarajan and C. Kakkar, "S-Band Single-Stage EDFA With 25-dB Gain Using Distributed ASE Suppression", IEEE Photonics Technol. Lett. 16 (11), (2004).

P. F. Wysocki, R. E. Tench, M. Andrejco, and D. DiGiovanni, in Proceedings of Optical Fiber Communications, OFC'97, pp. 127-129 Dallas, TX,1997.

P. F. Wysocky, in Proceedings of Optical Fiber Communications, OFC'98, pp. 97-99, San Jose, CA,1998.

P. Hajireza , S. D. Emami, C. L. Cham, D. Kumar, S. W. Harun and H. A. Abdul-Rashid" Linear All-fiber Temperature Sensor based on Macro-Bent Erbium Doped Fiber" ,Laser Physycs Letter, vol. 7, No. 10, pp. 739-742, 2010

P. Hajireza , S. D. Emami, S. Abbasizargaleh, S. W. Harun and H. A. Abdul-Rashid "Optimization of Gain flattened C-band EDFA using macro-bending" Laser Physics Letter, Vol. 20, No. 6, pp. 1–5 , 2010

P. Hajireza, S. D. Emami, S. Abbasizargaleh, S. W. Harun, D. Kumar and H. A. Abdul-Rashid, „Application of Macro-Bending for Flat and Broad Gain EDFA" Journal of Modern Optics, 2010

P. Hajireza, S. D. Emami, S. Abbasizargaleh, S. W. Harun, D. Kumar, and H. A. Abdul-Rashid "Temperature Insensitive Broad And Flat Gain C-Band Edfa Based On Macro-Bending", Amplifier" Progress In Electromagnetics Research C, Vol. 15, Pp. 37-48

P. Myslinski and J. Chrostowski, "Gaussian mode radius polynomials for modeling doped fiber amplifiers and lasers," Microwave OpticTechnology Letter, vol. 11 no. 2, pp. 61–64, 1996.

Parekhan M. Aljaff, and Banaz O. Rasheed "Design Optimization for Efficient Erbium-Doped Fiber Amplifiers" World Academy of Science, Engineering and Technology, 2008

S. A. Daud, S. D. Emami, K. S. Mohamed, H. A. Abdul-Rashid, S. W. Harun, H. Ahmad, M. R. Mokhtar, Z. Yusoff' and F. A. Rahman, in Proceedings of IEEE Conference on Photonics Global Institute of Electrical and Electronics Engineers, Singapore, pp. 1-3, 2008.

S. A. Daud, S. D. Emami, K. S. Mohamed, N. M. Yusoff, L. Aminudin, H. A. Abdul-Rashid, S. W. Harun, H. Ahmad, M. R. Mokhtar, Z. Yusoff and F. A. Rahman, "Gain and Noise Figure Improvements In a Shorter Wavelength Region of EDFA Using A Macro-Bending Approach", vol.18, no. 11, pp. 1362-1364, 2008.

S. D. Emami, P. Hajireza, F. Abd-RahmanF. Abd-Rahman, H. Ahmad, "wide-band hybrid amplifier operating in s-band region" Progress In Electromagnetics Research, PIER pp. 301, 313, 2010.

S. D. Emami, S. W. Harun, F. Abd-Rahman, H. A. Abdul-Rashid, S. A. Daud, and H. Ahmad "Optimization of the 1050nm Pump Power and Fiber Length in Single-pass and Double-pass Thulium Doped Fiber Amplifier" PIERB 14 , pp. 431-448, 2009.

S. W. Harun, K. Dimyati, K. K. Jayapalan, and H. Ahmad, "An Overview on S-Band Erbium-Doped Fiber Amplifier," Laser Physics Letter, vol. 4 , pp. 10–15, 2006.

S.K. Yun et al., Dynamic erbium-doped fibre amplifier based on active gain flattening with fibre acoustooptic tunable filter, IEEE Photonics Technology Letter. Vol. 11 pp. 1229–1231,1999.

S.W. Harun, , S.D. Emami, F. Abd Rahman, S.Z. Muhd-Yassin, M.K. Abd-Rahman, and H. Ahmad "Multiwavelength Brillouin/Erbium Ytterbium fiber laser" Laser Physiccs Journal, pp: 1-3,2007

S.W. Harun, R. Parvizi, X.S. Cheng, A. Parvizi, S.D. Emami, H. Arof and H. Ahmad ' Experimental and theoretical studies on a double-pass C-band bismuth-based erbium-doped fiber amplifier 'Optics & Laser Technology, vol. 42, no. 5, pp. 790-793 , July 2010

T. Pfeiffer, H. Bulow "Analytical Gain Equation for Erbium-Doped Fiber Amplifiers Including Mode Field Profiles and Dopant Distribution" IEEE Photonics Technology Letters, vol. 4, no. 5 , May 1992

Uh-Chan Ryu, K. Oh, W. Shin, U. C. Paek, IEEE Journal of Quantum Electronic vol. 38, pp. 149-161, 2002.

8

The Composition Effect on the Dynamics of Electrons in Sb-Based QD-SOAs

B. Al-Nashy[1] and Amin H. Al-Khursan[2]
[1] *Science College, Missan University, Missan,*
[2]*Nassiriya Nanotechnology Research Laboratory (NNRL), Science College, Thi-Qar University, Nassiriya, Iraq*

1. Introduction

In 1961, it is suggested that stimulated emission can occur in semiconductors by recombination of carriers that are injected across a p-n junction. This is the backbone in the concept of semiconductor lasers and amplifiers [1]. This concept is connected with the "one dimensional electron in a box", a problem discussed by quantum mechanics text books, to generate a new field of confined semiconductor structures known thereafter as quantum-well (QW), quantum-wire (QWi) or quantum dot (QD) structures where the carriers are confined in semiconductor in one, two, or three directions, respectively [2]. Semiconductor optical amplifiers (SOAs) and lasers performance may be substantially improved by using the QD-SOAs characterized by a low threshold current density, high saturation power, broad gain bandwidth and week temperature dependence as compared to bulk and multi-quantum well devices [3]. QD active region results from reducing the size of conventional (bulk) crystal to a nanometer size scale in all the crystal directions. This results in a complete quantization of energy states. The density of states becomes comparable to a delta-function. In order to cover losses of the waveguide of the SOA, QDs are then grown by a large number of dots grown on a WL which is already in the form of QW layer, see Fig. 1.
In this chapter, we examine the effect of changing composition of the layers constructing the Antimony (Sb)- based quantum dot semiconductor optical amplifiers (QD-SOAs). We start-off in sections 2-3 a general description of QD-SOA and the importance of SOAs, especially Sb-based SOAs, in solving problems. Manufacture imperfection in the shape of QDs are taken into account in the gain calculations in section 4, where they are represented by an inhomogeneous function. Along with the gain description in this section, a global Fermi-function is used where the states in the barrier layer, wetting layer (WL) and ground and excited states of the dots are included in the calculations of Fermi-energy. This mathematical formulation coincides with our subject of study since we would like to see the effect of all the QD-SOA layers. QD-SOA is modeled in section 5 using rate equations (REs) model for the barrier layer which is assumed to be in the form of separate confinement heterostructure layer (SCH), wetting layer (WL), ground state (GS) and excited state (ES) in the QD region of Sb-based structures and then solved numerically. A 3 REs model is discussed first where the

barrier SCH layer is neglected in this model, as done in many literatures, then a 4 REs model is described, where the SCH layer is included. Sb-based QD-SOA structures, the matter of study, are described in detail both in the shape and composition in section 6. Results of the calculations from these models are described in detail in section 7 where the material confinement is examined through the changing of Sb-composition in the QD, WL and SCH layers and it is shown to affects QD-SOA gain and dynamics. Changing WL composition affects the dynamics of ES and GS. InSb dots are shown to be more appropriate for inline static amplification. Results from 3 REs model overestimates the carrier dynamics in comparison with 4 REs calculations. Thus, the SCH barrier layer must be included in the RE models for convenient description of the processes in the QD-SOA. Our results supports the importance of inclusion global quasi-Fermi energy in explaining the results. Finally, section 8 concludes the main finding from this chapter.

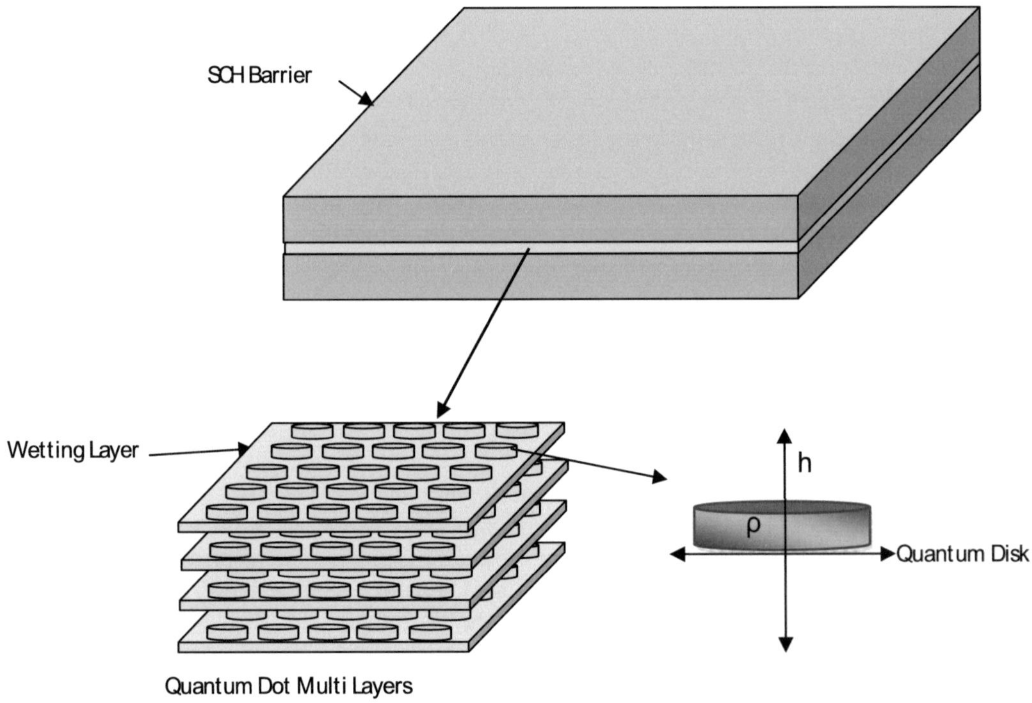

Fig 1. Schematic illustration of the QD active layer which consists of an arrays of Quantum Disks grown on an wetting layer covered by SCH barrier.

2. Semiconductor optical amplifier

Transmitters, receivers, electronic switches and routers limits the capacity of optical communication systems that can exceed 10 Tb/s by the speed of their electronic components [3]. Optical fibers suffers from attenuation (which limits transmission distance) and dispersion (leading to an increase in the system bit error rate, BER). 3R method (reshaping-retiming-retransmission) is then used to regenerate the optical signal in optical fibers. It has a number of disadvantages. This involves low optical and electrical transparencies in addition to network unreliability. These limitations may be overcome by using SOAs. The electronic components can be replaced by ultrafast all-optical signal processing components. SOAs are among the most promising candidates for all-optical signal processing devices due to their high speed capability, low switching energy, compactness, and optical integration compatibility. In optical fibers, as the in-line amplifier has only to carry out one function (amplification of the input signal) compared to full regeneration, it is intrinsically a more reliable and less expensive device. Optical amplifiers can also be a useful as power boosters, for example to compensate for splitting losses in optical distribution networks. It can be used as optical preamplifiers to improve receiver sensitivity [3,4].

A semiconductor optical amplifier (SOA) is basically a semiconductor laser (gain medium) with a low feedback mechanism and whose excited carrier amplifies an incident signal but do not generate their own coherent signal. Thus, it operated as a broadband single-pass device for amplification. Each SOA requires some form of external power (a current or optical source) to provide the energy for amplification. An electrical current inverts the medium, by transferring electrons from the valence to the conduction band, thereby producing spontaneous emission and the potential for stimulated emission yields the signal gain. The first studies of SOAs were carried out around the time of the invention of the semiconductor laser in the 1960's. In the 1970's Zeidler and Personick carried out early work on SOAs. Research on SOA device design and modeling gets a lot of importance in 1980's especially for AlGaAs SOAs operating at (830 nm) wavelength and InGaAsP/InP SOAs operating in the (1300-1550 nm) region. In 1989 SOAs designed as devices uses a symmetrical waveguide structures giving much reduced polarization sensitivities [5].

3. Sb-based QD-SOA

Mid-to-far-infrared wavelength range (2000-30000 nm) has a variety of applications including remote sensing, medical diagnostic, free-space optical communications, atmospheric pollution monitoring, chemical sensing, thermal imaging, high power and mid-infrared light sources [6]. Although HgCdTe material is already used to build devices working at these ranges, it suffers from slow response, high-power dissipation and undesired noise [7]. III-Sb based devices appear as a counterpart, since it is offering several advantages such as wide spectral range, low electron effective mass and high mobility at room temperature [8]. Although its research begins since the 70^{s} of the past century, Sb-based devices still need a lot of length of study especially when the active region is in the form of QD nanostructures. Using QD active region in the Sb-based devices adds the advantages resulting from QD nanostructures to that expected from the Sb-based infrared devices.

QD-SOAs can be used in the building blocks of many photonic devices. Its response is controlled by their dynamic behavior which can be studied by the use of a three-level system of REs to describe QD-SOA states for ground state (GS), excited state (ES) and WL state, which is considered as a QW and then can be approximated by a single state. While quasi-Fermi levels in the conduction and valence subbands are typically calculated from the surface carrier density per QD layer, now in some recent researches, both WL and barrier layer are included [9]. The barrier layer considered here is assumed to be in the form of a separate confinement heterostructure layer (SCH). This layer used in semiconductor structures to assure carrier confinement, where the SCH layer must have a high bandgap compared to other layers constructing the semiconductor device. Accordingly, the barrier layer becomes included in the gain calculations. Due to this, we begin in our laboratory

(NNRL) a series of studies for some of the characteristics of Sb-based QD devices [10]. Here, the carrier dynamics in III-Sb based QD-SOAs are considered using four-level REs system. QD, WL and SCH barrier compositions are examined to specify their characteristics. The results then, compared with those of three-level REs show the importance of including the barrier layer in the QD SOAs calculations. Because of the much larger effective mass of holes and lower quantization energies of the QD levels in the valence band, electrons behavior limits the carrier dynamics while holes in the valence band are assumed to be in quasi-thermal equilibrium at all times [11]. Thus, we determine carrier dynamics here by the relaxation of electrons in the conduction band only.

4. Inhomogeneous gain model with global quasi-fermi levels

Actually, QDs have imperfections in shape and randomly distributed on the substrate. These QDs emit photons at slightly different energies which results in an inhomogeneous broadening [11]. In our model, a Gaussian distribution function is used to describe inhomogeneous broadening. Accordingly, the linear gain of QD structures can be written as [9, 12]

$$g^{(1)}(\hbar w)=c_0\sum_i\int_{-\infty}^{\infty}dE'\left|M_{env}\right|^2\left|e^{\wedge}.p_{cv}\right|^2D(E')L(E',\hbar w)[f_c(E',F_c)-f_v(E',F_v)] \quad (1)$$

where $c_0=\dfrac{\pi e^2}{m_0^2\varepsilon_0 c n_b\omega}$, m_0 is the free electron mass, ε_0 is the permittivity of free space, c is the speed of light in free space, n_b is the background refractive index of the material, ω is the optical angular frequency of the injected optical signal and E' is the optical transition energy. $\left|M_{env}\right|^2$ is the envelope function overlap between the QD electron and hole states. The term $[\left|e^{\wedge}.p_{cv}\right|^2=\frac{3}{2}(\frac{m_0}{6})E_p]$ is the momentum matrix element for electron-heavy hole transition energy in TE polarization, E_p is the optical matrix energy parameter. The Lorentzian line shape function $L(E')$ for gain is defined by

$$L(E',\hbar w)=\frac{\frac{\hbar\gamma_{cv}}{\pi}}{(E'-\hbar\omega)^2+(\hbar\gamma_{cv})^2} \quad (2)$$

With $\gamma_{cv}(=1/\tau_{in})$ is the intraband scattering rate. $D(E')$ is the inhomogeneous density of states of the self-assembled QD and is expressed as [13]

$$D(E')=\frac{\mu_i}{V_{dot}}\frac{1}{\sqrt{2\pi\sigma^2}}\exp\left(\frac{-(E'-E^i_{\max})^2}{2\sigma^2}\right) \quad (3)$$

where μ_i is the degeneracy of the ith state of a QD, $\mu_{GS}=2,\mu_{ES}=4$ for the ground state and the excited state, respectively. V_{dot} is the effective volume of the QDs. σ is the spectral variance of the QD distribution and $E^i_{\max}$ is the transition energy at the maximum of QD distribution of the ith optical transition. The terms f_c and f_v are the respective quasi-Fermi level distribution functions for the conduction and valence bands, respectively. Recent researches [8] uses global states to describe the global quasi-Fermi levels F_c and F_v in the conduction and valence bands where the contributions to the Fermi-levels from the barrier layer and WL are included in addition to that from QDs. They are determined from the surface carrier density per QD layer by the following relations [13]

$$\begin{aligned} n_{2D} = {} & N_D \sum_i \frac{s^i}{\sqrt{2\pi\sigma_e^2}} \int e^{-\left(E_c'-E_{ci}^D\right)/2\sigma_e^2} f_c\left(E_c', F_c\right) dE_c' \\ & + \sum_l \frac{m_e^W K_B T}{\pi\hbar^2} \ln\left(1 + e^{\left(F_c - E_{cl}^W\right)/K_B T}\right) \\ & + H_b \int \frac{1}{2\pi^2}\left(\frac{2m_e^B}{\hbar^2}\right)^{3/2} \sqrt{E_c'-E_c^B}\, f_c\left(E_c', F_c\right) dE_c' \end{aligned} \tag{4}$$

$$\begin{aligned} p_{2D} = {} & N_D \sum_i \frac{s^i}{\sqrt{2\pi\sigma_h^2}} \int e^{-\left(E_h' - E_{hj}^D\right)/2\sigma_h^2} f_v\left(E_h', F_v\right) dE_h' \\ & + \sum_m \frac{m_h^w K_B T}{\pi\hbar^2} ln\left(1 + e^{\left(F_v - E_{hm}^w\right)/K_B T}\right) \\ & + H_b \int \frac{1}{2\pi^2}\left(\frac{2m_h^B}{\hbar^2}\right)^{3/2} \sqrt{E_h^B - E_h'}\, f_v\left(E_h', F_v\right) dE_h' \end{aligned} \tag{5}$$

where n_{2D} and p_{2D} are the surface densities of electrons and holes per QD layer, respectively. E_{ci}^D and E_{hi}^D represents the respective confined QD states in the conduction and valence bands. σ_e and σ_h are the spectral variance of the QD electron and heavy-hole distributions, respectively. The term E_c' is the energy in the conduction band and E_h' is the energy in the valence band. Heavy hole subbands are included, only, in the calculations of valence subbands since light hole subbands are deep and can be neglected. k_B is the Boltzmann constant and T is the absolute temperature. The terms m_e^W (m_h^W) and E_{el}^W (E_{hm}^W) are the effective electron (hole) mass and the subband edge energy of the conduction (valence) band of WL. The term H_b is the thickness of SCH barrier. The terms m_e^B (m_h^B) and E_c^B (E_h^B) are the electron (hole) mass and the conduction (valence) band edge energy of barrier layer.

5. Rate equations of Sb-based QD-SOAs

5.1 Three rate equations model

QD-SOA characteristics can be studied using REs system constructed from three equations which describe the dynamics in WL, GS and ES of QD layer. The QD inhomogeneity due to fabrication imperfections is introduced through gain as discussed in section 3 above. Examples of three REs system can be found in [11, 14]. In the three REs system, carriers are assumed to be injected with current density, *J*, in WL where the barrier layer effect is neglected. In WL, carriers are recombining at a rate $(1/\tau_{wR})$, relax at a rate $(1/\tau_{w2})$ to the ES. ES relaxes quickly to GS at rate $(1/\tau_{21})$, thus it does not contribute to gain. The escape component from ES to WL is $(1/\tau_{2w})$ and from GS to ES is $(1/\tau_{12})$. Both GS and ES are assumed to be recombining at the same rate $(1/\tau_{1R})$, see Fig. 2. The Pauli blockings $(1-h)$ and $(1-f)$ are taken for the nonempty ES and GS, respectively. For example, if ES is empty, then (h=0) and carriers can occupy the state, if it is fully occupied, then (h=1) and no further carriers can occupy it. The above described processes are represented by the following REs model for the carrier density N_w in the WL, occupation probabilities h and f in the ES and GS, respectively

$$\frac{\partial N_w}{\partial t} = \frac{J}{qL_w} - \frac{N_w(1-h)}{\tau_{w2}} + \frac{N_w h}{\tau_{2w}} - \frac{N_w}{\tau_{wR}} \tag{6}$$

$$\frac{\partial h}{\partial t} = \frac{N_w L_w(1-h)}{N_Q \tau_{w2}} - \frac{N_w L_w h}{N_Q \tau_{2w}} - \frac{(1-f)h}{\tau_{21}} + \frac{f(1-h)}{\tau_{12}} \tag{7}$$

$$\frac{\partial f}{\partial t} = \frac{(1-f)h}{\tau_{21}} - \frac{f(1-h)}{\tau_{12}} - \frac{f^2}{\tau_{1R}} - S_{av}\Gamma \frac{g_{\max}(\hbar w_p)}{N_Q}(2f-1)L\frac{c}{n_g} \tag{8}$$

where q is the electron charge, L_w is the effective thickness of the active layer, Γ is the optical confinement factor and N_Q is the surface density of QDs. The average signal photon density S_{av} is given by [14]

$$S_{av} = \frac{g_s P_{in} n_g}{\hbar w_p L_w D L g_{\max} c} \tag{9}$$

P_{in} is the input signal power to the SOA, n_g is the group refractive index, $\hbar$ is the normalized Plank's constant, w_p is the peak frequency, D is the strip width, L is the cavity length, c is the free space light speed, g_{max} is the peak material gain taken at the peak frequency. The input signal wavelength is assumed to be injected to the QD SOA at a peak wavelength of GS for each structure. The single-pass gain of the structure is given by [11]

$$g_S = \exp\left[\left(g_{\max}\Gamma - a_{\text{int}}\right)L\right] \tag{10}$$

where a_{int} is the loss coefficient. Although WL is assumed here to receive carriers by current injection or by escapes from QD ES, but the three RE models consider it as a common reservoir [10, 15]. From the architecture of these three RE model, ES works as a common reservoir since it receives carriers from the states above and below it (WL and GS, respectively). So, ES is referred to as a reservoir for GS [8]. This is contradicted with the experimentally evidenced that carriers stay long in WL and short in ES [16]. Due to this and other reasons, discussed below, four REs system is more appropriate.

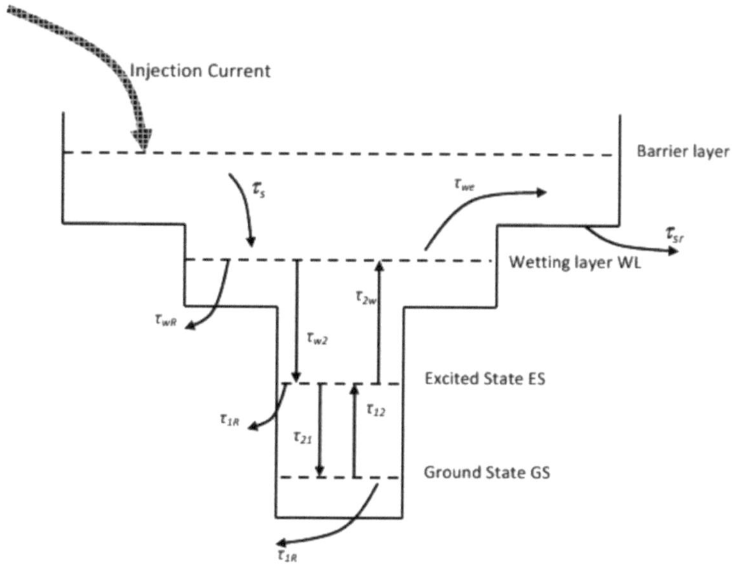

Fig. 2. Energy diagram of QD-SOA system.

5.2 Four rate equations model

Four REs system is used here to study Sb-based QD-SOAs and is built depending on the energy band diagram considered in Fig. 2. The advantages of using four rate equations system are twofold: 1) including of SCH barrier layer dynamics which is important in practice and 2) this RE model is more appropriate with the use of global Fermi-energy in the gain calculations which is described in section 2. In this four REs system, carriers are assumed to be injected with current density, J, in the SCH barrier layer then recombine at a rate $(1/\tau_{sr})$, relax at rate $(1/\tau_s)$ to the WL which works as a common reservoir for the carriers. The escape rate from WL to the SCH is $(1/\tau_{we})$. Other processes are described in section 5.1. The processes in the four REs system are represented by the following REs model which covers carrier density, N_s, in SCH barrier layer, in addition to: N_w, h and f which are covered also in the three REs model

$$\frac{\partial N_s}{\partial t} = \frac{J}{qL_w} - \frac{N_s}{\tau_s} - \frac{N_s}{\tau_{sr}} + \frac{N_w}{\tau_{we}} \tag{11}$$

$$\frac{\partial N_w}{\partial t} = \frac{N_s}{\tau_s} - \frac{N_w(1-h)}{\tau_{w2}} + \frac{N_w h}{\tau_{2w}} - \frac{N_w}{\tau_{wR}} - \frac{N_w}{\tau_{we}} \tag{12}$$

$$\frac{\partial h}{\partial t} = \frac{N_w L_w (1-h)}{N_Q \tau_{w2}} - \frac{N_w L_w h}{N_Q \tau_{2w}} - \frac{(1-f)h}{\tau_{21}} + \frac{f(1-h)}{\tau_{12}} \tag{13}$$

$$\frac{\partial f}{\partial t} = \frac{(1-f)h}{\tau_{21}} - \frac{f(1-h)}{\tau_{12}} - \frac{f^2}{\tau_{1R}} - S_{av}\Gamma \frac{g_{\max}(\hbar w_p)}{N_Q}(2f-1)L\frac{c}{n_g} \tag{14}$$

Quasi-thermal equilibrium is assumed between states. To ensure this convergence, the carrier escape times are related to the carrier capture times as follows [17]

$$\tau_{12} = \tau_{d0}(\mu_{GS} / \mu_{ES})e^{([E_{ES}-E_{GS}]/k_B T)} \quad (15)$$

$$\tau_{2w} = \tau_{c0}(\mu_{ES} N_Q / \rho_{weff})e^{([E_{wl}-E_{ES}]/k_B T)} \quad (16)$$

$$\tau_{we} = \tau_s(\rho_{weff} N_{QD} / \rho_{SCH} H_b)e^{(\Delta E_{SCH,wl}/k_B T)} \quad (17)$$

GS and ES QD energy levels are denoted by E_{GS}, E_{ES}. ρ_{weff} is the density of states per unit area in the WL and ρ_{SCH} is the density of states per unit volume in the SCH. They are given by $\rho_{weff} = (m_{ewl} k_B T / \pi\hbar^2)$ and $\rho_{SCH} = 2(2m_{eSCH}\pi k_B T / \hbar^2)^{3/2}$. N_{QD} is the number of QD layers and H_b is the total thickness of the SCH. $\Delta E_{SCH,wl}$ is the energy difference between SCH and WL band edge energies. The capture times τ_{c0} and τ_{d0} are the average capture time from the WL to the ES and from the ES to the GS with the hypothesis that the final state is empty.

6. Sb-based QD-SOA structures

Sb-based QD semiconductors suffer from the difficulty of QD growth methods, like the self-assembled growth method, due to kinetic effects. Sb-based QD crystals growth hinders due to large mismatch (>7 or 8%) with GaAs semiconductors. Now it is possible to grow GaSb QDs on GaAs substrate [18]. In addition, an interfacial misfit array method can be used now to grow Sb-based layers on a GaAs platform although the (8%) lattice mismatch between GaAs and GaSb [19].

Accordingly, we choose a large number of structures, it is classified here in a five types, as in Table 1. In these structures, we change the composition of QD (structure Nos. 1, 4 and 5), WL (No. 2) or barrier layer (No. 3). This is to examine the effect of these layers on the QD-SOA and also to specify the possible spectral ranges in the Sb-based QD structures. In all of these structures there is a lattice mismatch between layers (dot and WL or WL and barrier) not exceeds (6%) and in some cases its value is very small. This is to intend these structures to the ease of QD growing. QD energy levels are calculated using parabolic band quantum-disc model [20] where the dots are assumed to be in the form of a disc with radius $(14 nm)$ and height $(h = 2nm)$, unless states otherwise. Quantum well WL thickness is taken as (9 nm). The accuracy of the quantum disc model is checked by a comparison with the experimental data and numerical methods [20, 21]. The parameters used in the calculations of Sb-based structures are stated in Tables 2 and 3. Thus, it is adequate to calculate energy levels without time consumption. An example of QD energy levels calculated using parabolic band model is shown in Fig. 3 for $InAs_{0.1}Sb_{0.9}$ / $GaAs_{.1}Sb_{.9}$/ $Al_{.1}Ga_{.9}As$ QD-SOA. In the calculation of quasi-Fermi energy Eqs. (3) and (4) are used. Gain is then calculated, using Eq.1, and its peak value and peak wavelength for each structure is specified.

No.	Structure	No.	Structure
1	$InAs_xSb_{1-x}/GaAs_{0.1}Sb_{0.9}/Al_{0.1}Ga_{0.9}As$	4	$GaSb/GaAs_{0.7}Sb_{0.3}/GaAs$
2	$In_{0.1}AsSb_{0.9}/GaAs_xSb_{1-x}/Al_{.1}Ga_{.9}As$	5	$InSb/GaAs_{0.7}Sb_{0.3}/GaAs$
3	$In_{0.1}AsSb_{0.9}/GaAs_{0.1}.Sb_{0.9}/Al_xGa_{1-x}As$		

Table 1. QD-SOA structures studied.

Parameter	Symbol (Unit)	InSb	GaSb	InAs	GaAs
Bandgap energy	$E_g\ (eV)$	0.17	0.73	0.36	1.424
Electron Effective Mass	m_e^* / m_0	0.0145	0.044	0.022	0.065
Heavy-hole Effective mass	m_{hh}^* / m_0	0.44	0.33	0.41	0.45
Refractive index	n	4	3.82	3.52	3.65

Table 2. Binary structure parameters [22].

Structure	Relation
$InAs_xSb_{1-x}$	$E_g(InAs_xSb_{1-x}) = 0.18 - 0.41x + 0.58x^2$ $m_i(ABC) = m_i(AC) + m_i(BC)$
$GaAs_xSb_{1-x}$	$E_g(GaAs_xSb_{1-x}) = 0.726 - 0.502x + 1.2x^2$ $m_i(ABC) = m_i(AC) + m_i(BC)$
$Al_xGa_{1-x}As$	$E_g(Al_xGa_{1-x}As) = 1.242 + 1.247x$ $m_i(ABC) = m_i(AC) + m_i(BC)$

Table 3. The relation used to calculate structure parameters (bandgap E_g and effective mass m_i). Note that the subscript (i) with m_i refers to conduction or valence band effective masses [22, 23].

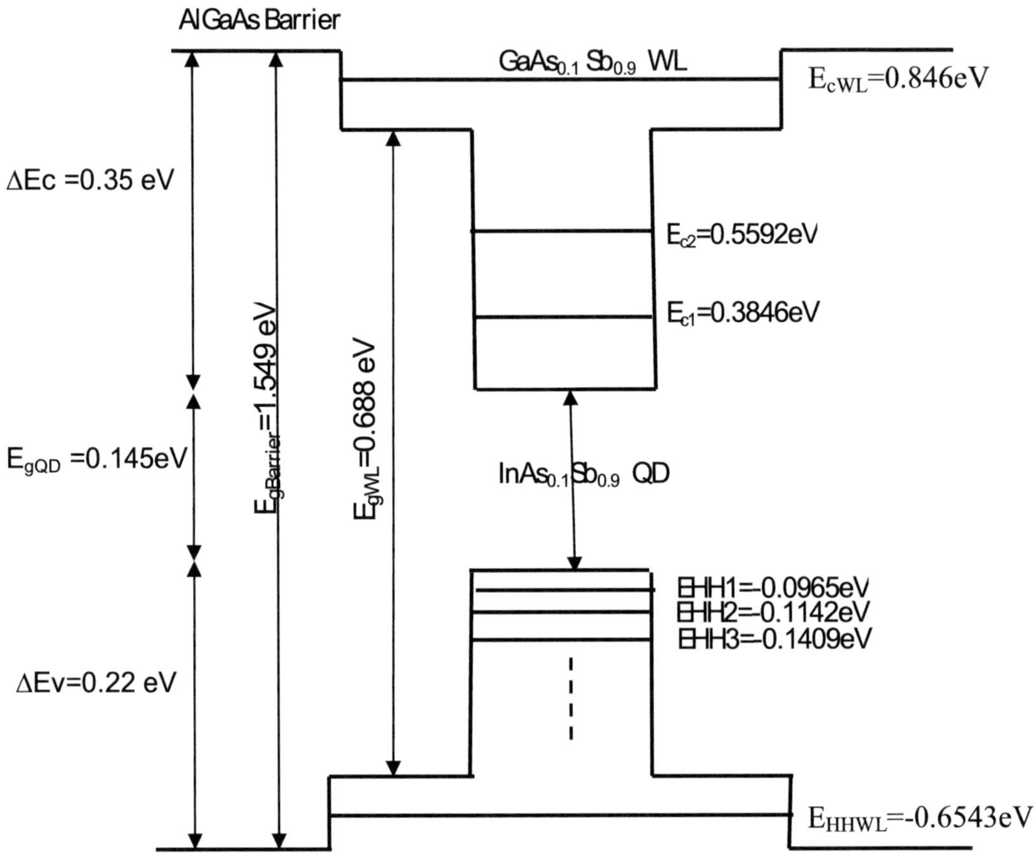

Fig. 3. Energy band diagram of an $InAs_{0.1}Sb_{0.9}/GaAs_{0.1}Sb_{0.9}/Al_{0.1}Ga_{0.9}As$ QD-SOA structure.

7. Simulation results and discussion

7.1 QD-SOA gain

The REs system are solved numerically to see the gain change with input power P_{in}, and to examine the dynamic effects. The parameters used in the numerical calculations are listed in Table 4. Fig. 4 shows input power-gain curves for Sb-based QD-SOAs studied. Fig. 4 (a) shows the effect of changing Sb mole fraction in the QD layer. The effect of WL, barrier and QD layers composition on gain is shown in Fig. 4 (b), (c), (d). Curve arrangement in these figures coincides with that in [12]. While both QD and WL composition are shown to give a respected changes in gain, the contribution due to changing Sb-composition in the barrier layer is minor as in Fig. 4 (c). The overall behavior of Fig. 4 is the same, gain saturates at low input power, then it declines at high input powers. This can be attributed to the carrier depletion in the QDs where the output power begins to increase. In Fig. 4 (d), InSb QD-SOA gives a higher gain than that obtained from GaSb QD-SOA. Comparing Fig. 4 (a) and (d) shows the effect of QD composition. To deal with composition effect, we must refer to the material confinement effect: the difference between bandgap in the dot and WL. The higher difference gives more carrier confinement in the QD layer. InSb is known as a lower band

Parameter	Symbol	Value	Unit
Spontaneous radiative lifetime in QDs	τ_{1R}	0.4	ns
Spontaneous radiative lifetime in WL	τ_{wR}	1	ns
Carrier escapes time from ES to WL	τ_{c0}	1	ps
Carrier relaxation time from the two-dimensional WL to the ES	τ_{w2}	3	ps
Carrier relaxation time from GS to ES	τ_{12}	1.2	ps
Carrier relaxation time from ES to GS	τ_{d0}	7	ps
Diffusion time in the barrier layer	τ_s	6	ns
SCH recombination time	τ_{sr}	4.5	ns
Internal loss	α_{int}	2	cm^{-1}
Optical confinement factor	Γ	0.007	
Laser length	L	2000	μm
Strip . width	D	10	μm
The effective thickness of the active layer	L_w	0.1	μm
Injection current density	J	1.335	kA/cm^2

Table 4. Parameters used in the calculations [11, 17].

gap semiconductor. Thus, a higher gain is obtained from InSb than GaSb QD-SOAs due to the higher material confinement of the former (GaSb bandgap differ from InSb by ~0.56 eV). This reason is not enough to explain curves arrangement in Fig. 4 (a). Here gain curves are not arranged due to QD band gap where only very small differences result from varying Sb mole fraction in the QD region. Thus, in addition to material confinement, one must refer to the quantum confinement. Curves are arranged due to the gap ($E_{c1}+E_g+E_{v1}$) where E_{c1}, E_{v1} are the 1st conduction and valence subbands and E_g is the QD band gap. From Figs. 4 (a) and (d), the structure $InSb/GaAs_{0.7}Sb_{0.3}/GaAs$, is more appropriate for inline static amplification applications due to its maximum gain obtained and higher saturation power than $InAs_xSb_{1-x}/GaAs_{0.1}Sb_{0.9}/Al_{0.1}Ga_{0.9}As$ and GaSb QD-SOAs. The effect of QD size is shown in Fig. 4 (e) and (f). In Fig. 4 (e), it is shown that QDs with shorter height gives a higher gain. In Fig. 4 (f), QDs with longer radius gives higher gain.

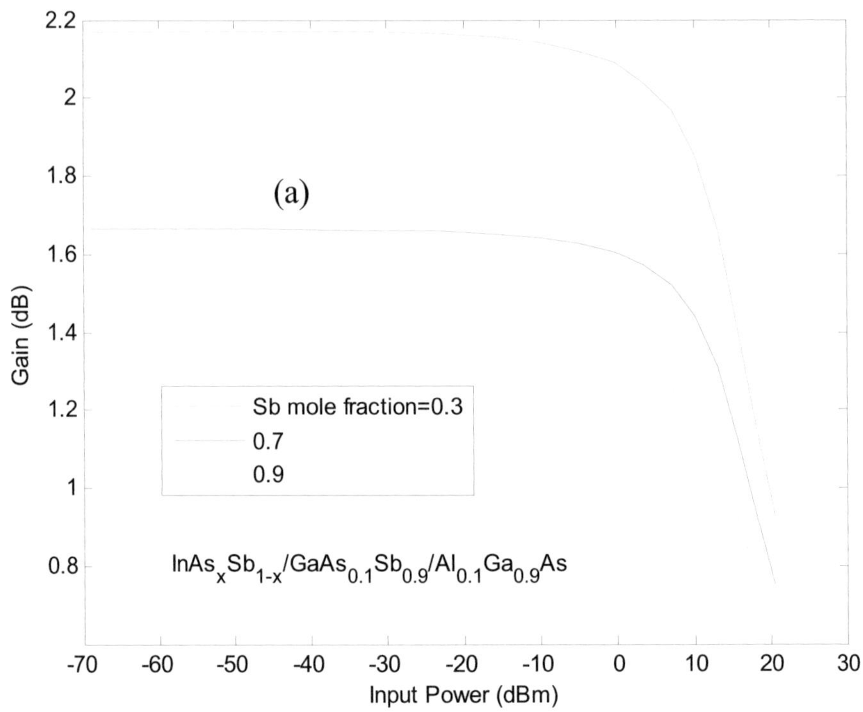

Fig. 4. (Continued)

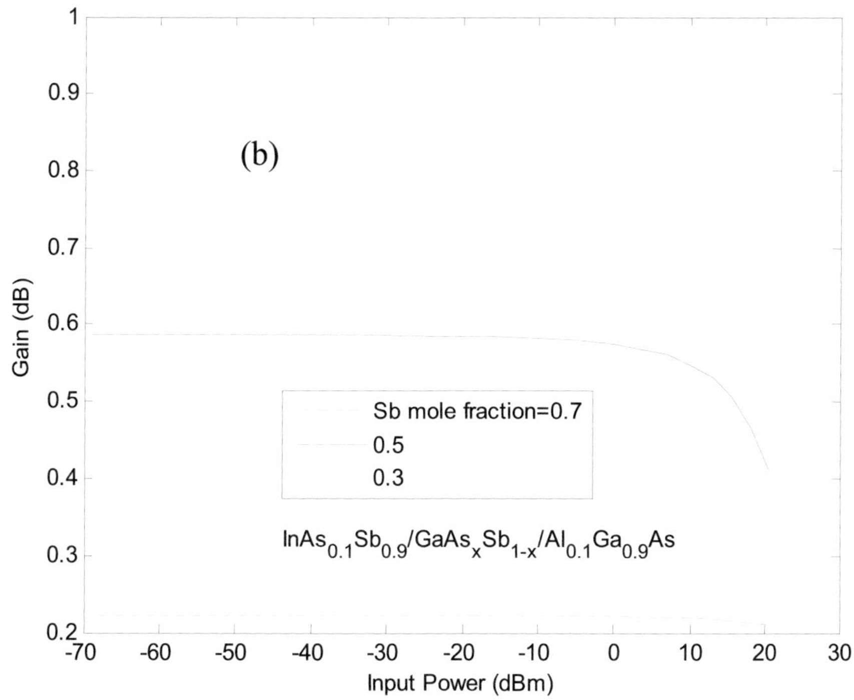

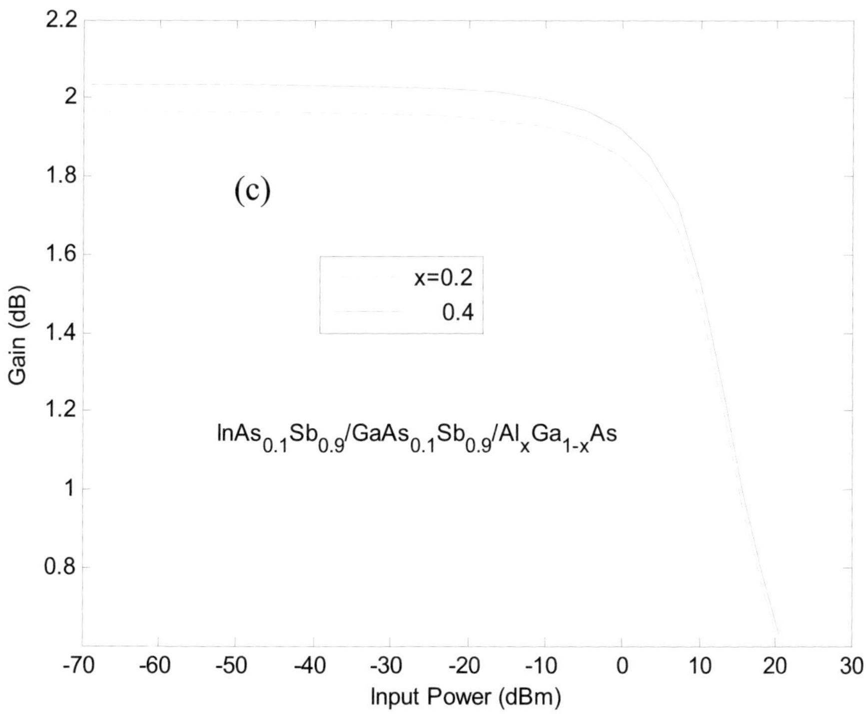

Fig. 4. (Continued)

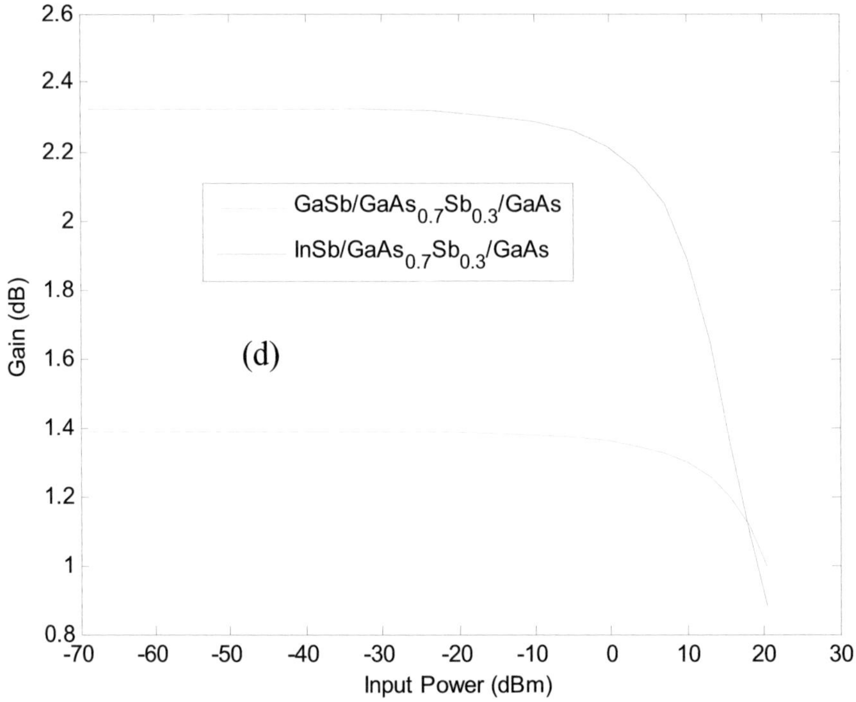

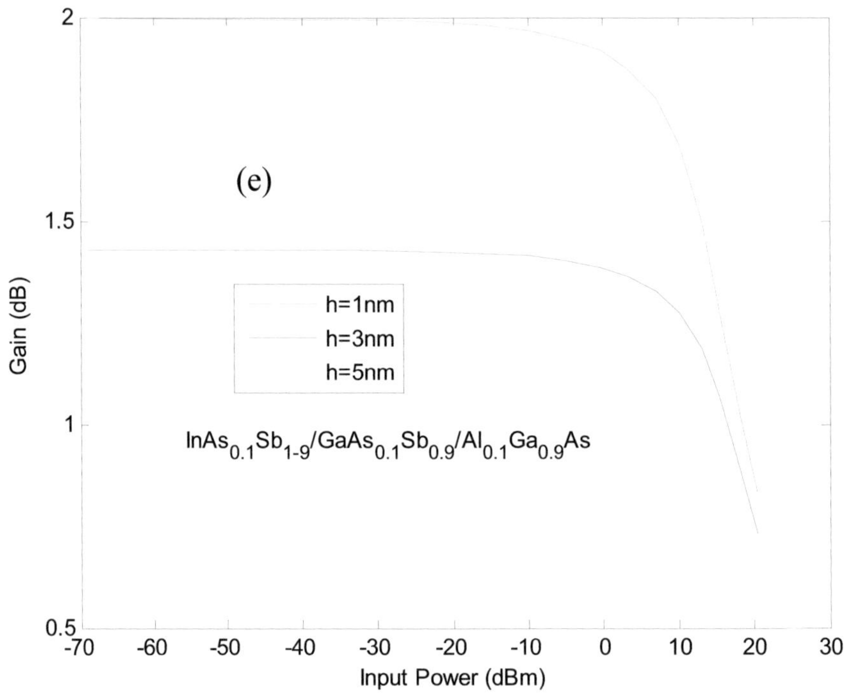

Fig. 4. (Continued)

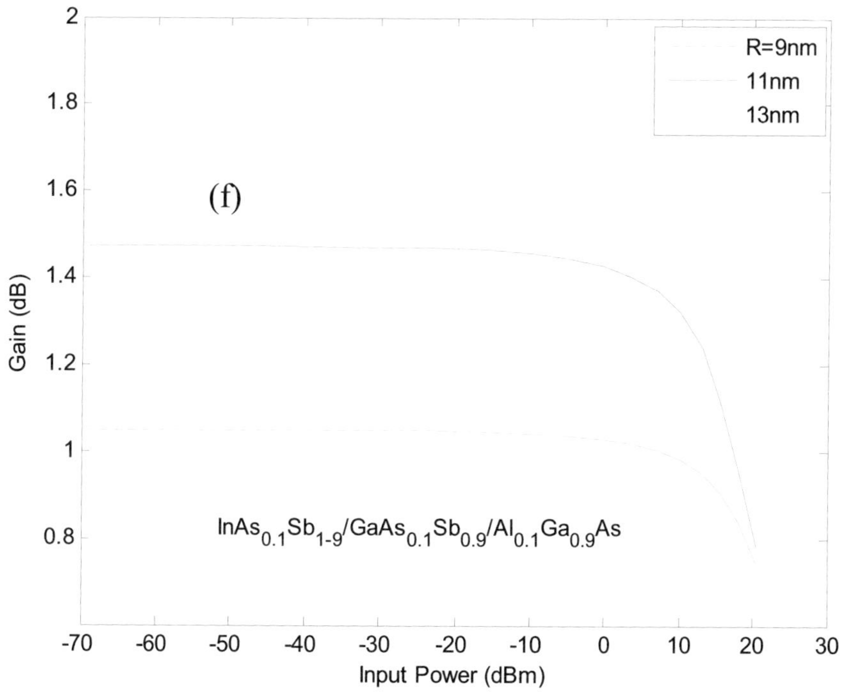

Fig. 4. Gain-input power relation at J=1.335 kA/cm[2] for the QD-SOAs: (a) $InAs_xSb_{1-x}/GaAs_{0.1}Sb_{0.9}/Al_{0.1}Ga_{0.9}$, (b) $InAs_{0.1}Sb_{0.9}/GaAs_xSb_{1-x}/Al_{0.1}Ga_{0.9}As$, (c) $InAs_{0.1}Sb_{0.9}/GaAs_{0.1}Sb_{0.9}/ Al_xGa_{1-x}As$, (d) GaSb and InSb. (e) The disc height and (f) the disc radius for $InAs_{.1}Sb_{.9}/GaAs_{.1}Sb_{.9}/Al_{.1}Ga_{.9}As$ QD-SOA is shown.

7.2 Dynamical effects

Figure 5 (a) shows the carrier density dynamics in the barrier layer N_s of $InAs_xSb_{1-x}/GaAs_{.1}Sb_{.9}/Al_{.1}Ga_{.9}As$ QD-SOAs at three x-mole fraction values (x=0.3, 0.7 and 0.9 Sb-fraction). The energy subbands of the QD at each mole fraction are included through relaxation times (see Eqs. 15-17), thus a difference appears between curves. The carrier density in the barrier layer obtained here is in the range of ($10^{15}cm^{-3}$) which is near to the value of WL carrier density in [11] which assumes that carriers are injected directly to WL. N_s curves are arranged according to the shallower Fermi energy level in the barrier layer of the QD SOA structure. According to this, the barrier layer in the structure with (0.9) Sb mole fraction filled earlier, since its quasi-Fermi energy is shallower, then the structure with (0.3) and finally, the structure with (0.7) Sb mole fraction. Fig. 5 (b) shows the carrier density in the WL (N_W). While N_W obtained here is in the range of ($10^{12}cm^{-3}$), a (10^{14} - 10^{15} cm^{-3}) WL carrier density is obtained in [11] due to the neglect of barrier layer in that work. Here, N_W curves are arranged according to the quasi-Fermi energy in their wetting layers. Fig. 5 (c) and (d) shows ES and GS occupation probabilities for the structures $InAs_xSb_{1-x}/GaAs_{.1}Sb_{.9}/Al_{.1}Ga_{.9}As$ at Sb mole fractions (0.3, 0.7 and 0.9) where the same probability is obtained in these structures for each ES and GS. This is due to very small differences between their relaxation times, i.e. the inclusion of QD subbands energies, not so much effects, see Eqs. (15-17). In Fig. 6 (a), the effect of changing mole-fraction in the WL of $InAs_{.1}Sb_{.9}/GaAs_xSb_{1-x}/Al_{.1}Ga_{.9}As$ QD structure is studied. WL is shown to be saturated at

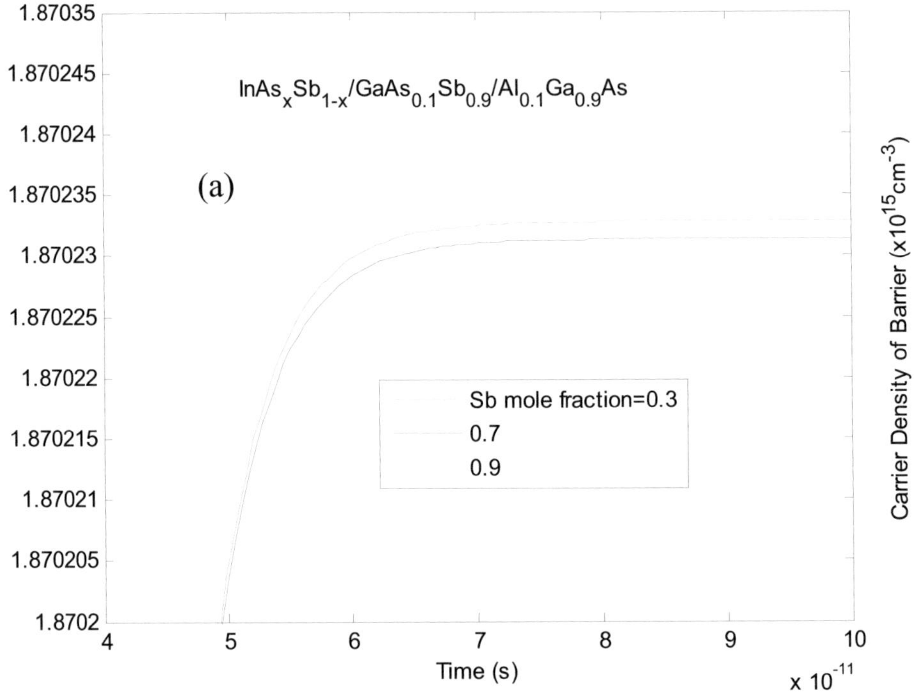

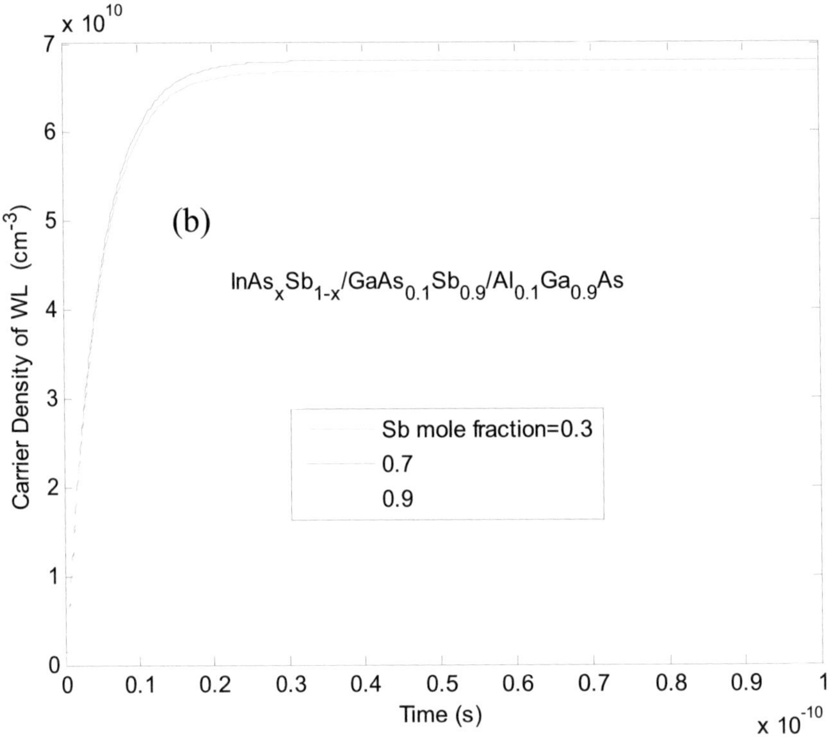

Fig. 5. (Continued)

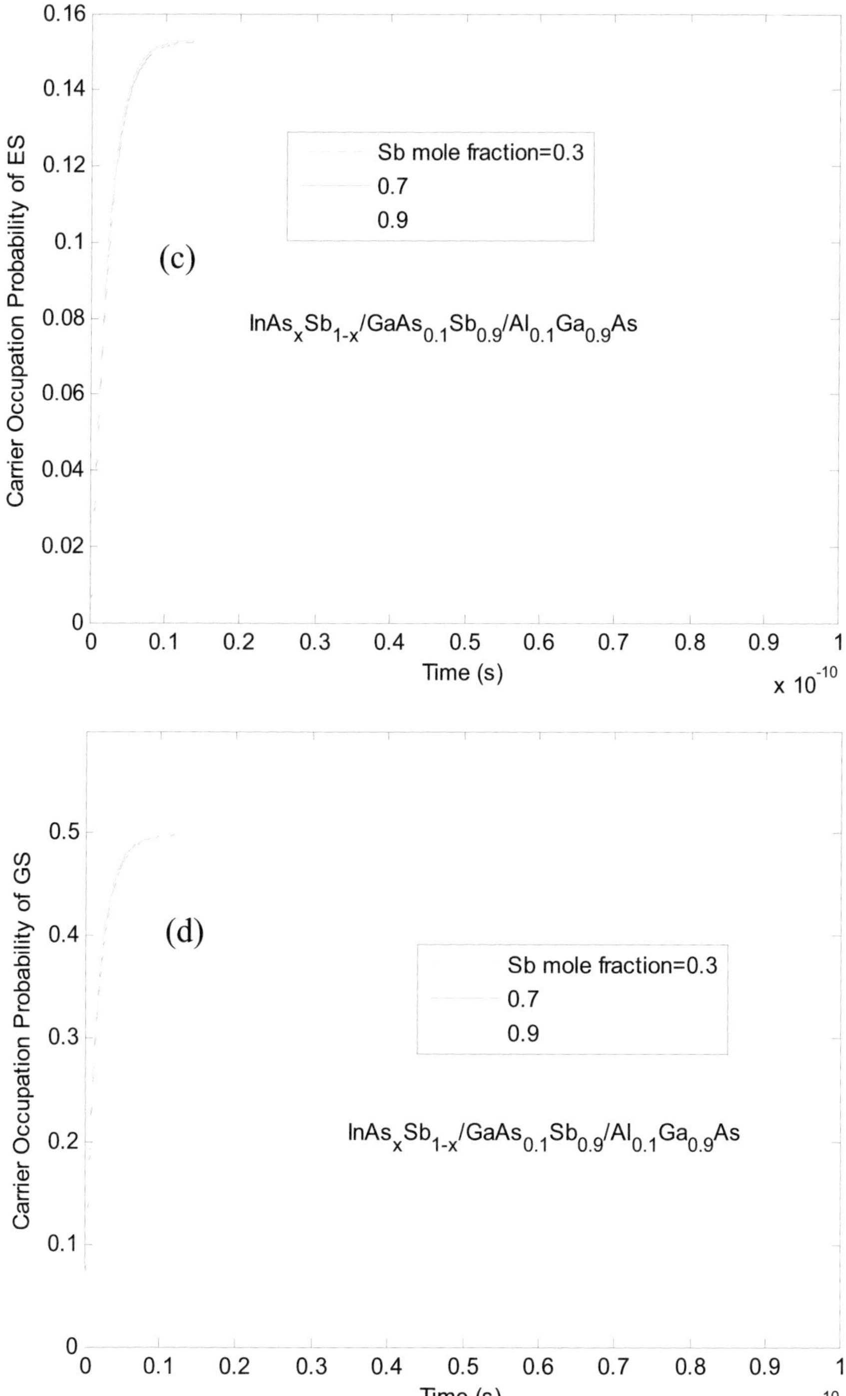

Fig. 5. Carrier density for (a) barrier and (b) WL. Then occupation probability for (c) ES, and (d) ES for $InAs_xSb_{1-x}/GaAs_{0.1}Sb_{0.9}/Al_{0.1}Ga_{0.9}As$ QD-SOAs.

higher carrier density for the structures with Sb mole fraction in WL (0.3 and 0.5). Here curves arrangement is appear according to shallower quasi-Fermi energy levels in the WL, where the effect of WL band gap energy is obvious in this arrangement. Fig. 6 (b) and (c) shows the dynamics of ES and GS occupation probabilities, respectively. Checking the parameters that arrange these curves shows that although the escape times to ES (τ_{12} and τ_{2w}) are very short for the structure No. 2 with Sb mole fraction (0.7), it saturates after other mole fractions (0.3 and 0.5) in these structures. This can be explained if one follows the energy difference between QD ES and WL energy level where this difference is greater for the structure with (0.7) Sb mole fraction in the WL. Also for GS one must refer to the difference between QD GS subband and WL energy level. In Fig. 7, the WL dynamics are shown for $In_{0.1}AsSb_{0.9}/GaAs_{0.1}Sb_{0.9}/Al_xGa_{1-x}As$ QD-SOAs. WL carrier density saturates at a higher value for the structure with (x=0.2) Al mole fraction in the barrier layer. Also the same reason for shallower quasi-Fermi energy in the WL can explain this arrangement. One can refer to effect of the main difference between structures here (energy gap of barrier layer E_{gb}) where a higher separation between N_W curves is compared to the effect of WL energy gap (E_{gw}) as shown in Fig. 6 (a). Occupation probabilities in GS and ES coincides for both (x=0.2 and 0.4) structures and thus they are not drown. In Figs. (5)-(6), although the faster rate of transition between ES and GS, but they are not always the faster one reaching steady state. Both SCH barrier and WL get steady state faster although there is a longer rate of transition between the barrier and WL. This is because of the limited dynamics for these layers so they are in the steady state earlier. At all cases, the GS reaches steady state faster

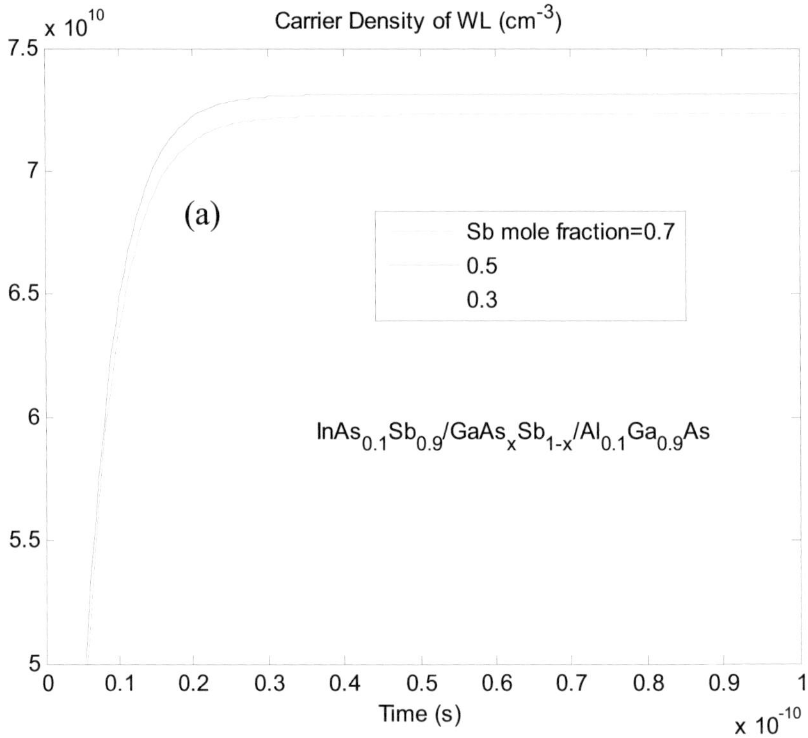

Fig. 6. (Continued)

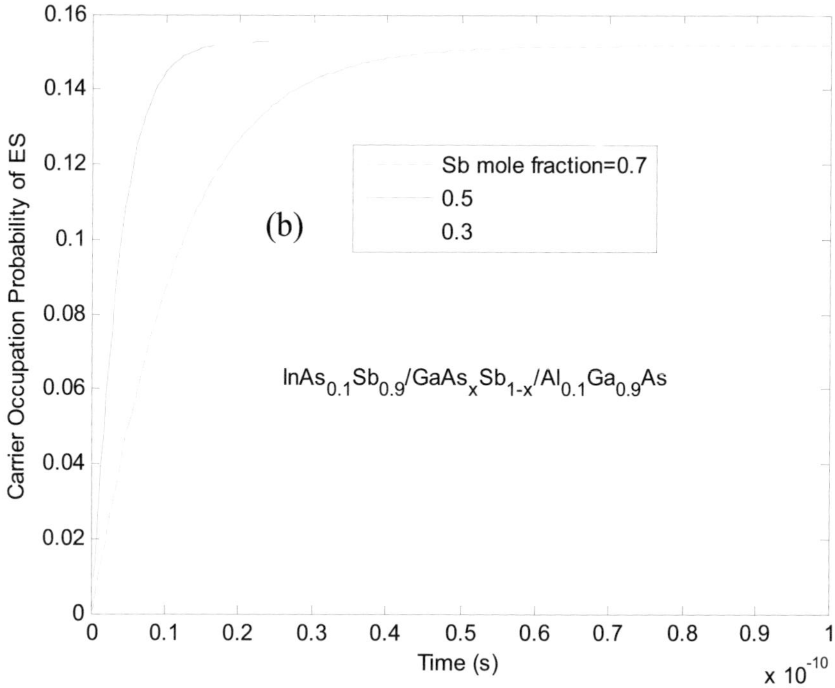

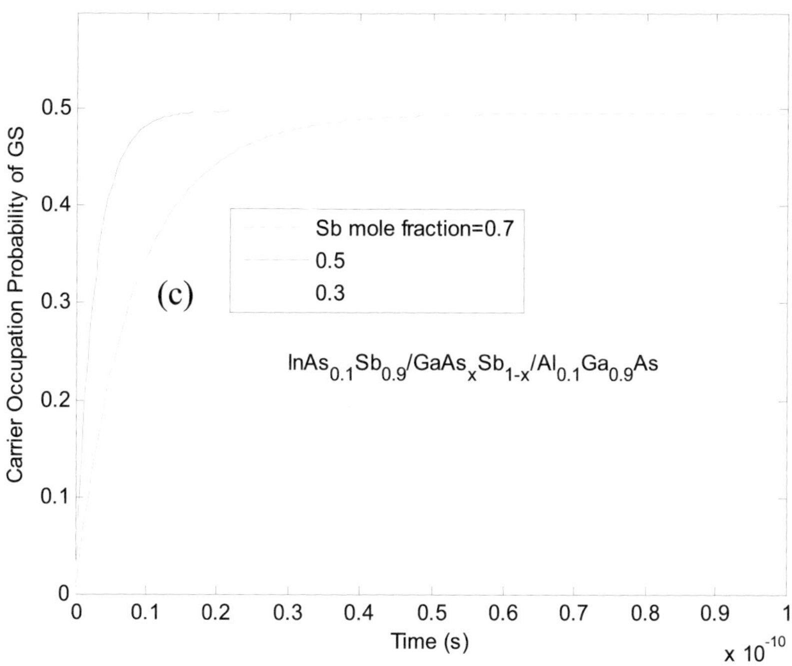

Fig. 6. (a) Carrier density for WL. Then occupation probability for: (b) ES and (c) GS for $InAs_{0.1}Sb_{0.9}/GaAs_xSb_{1-x}/ Al_{0.1}Ga_{0.9}As$ QD-SOAs

than ES since the relaxation between ES to GS is very fast. Finally, a comparison is done between three and four REs system for QD-SOA as shown in Fig. 8 (a), (b) and (c) for WL, ES and GS respectively, where a three REs system is shown to be an overestimates the dynamics due to neglecting the effect of barrier layer. So, this layer must be included in the REs system used to study QD systems.

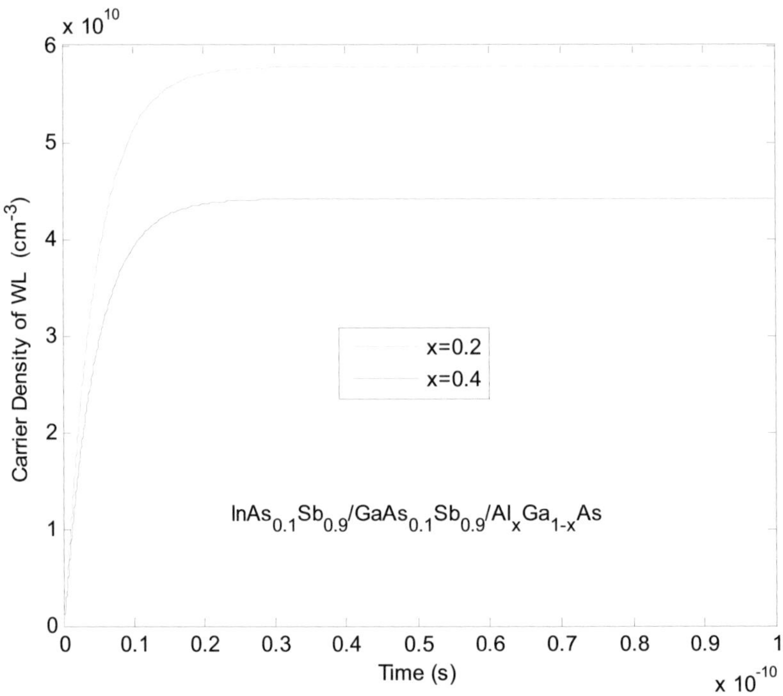

Fig. 7. Carrier density for WL for $InAs_{0.1}Sb_{0.9}/GaAs_{0.1}Sb_{0.9}/Al_xGa_{1-x}As$ QD-SOAs.

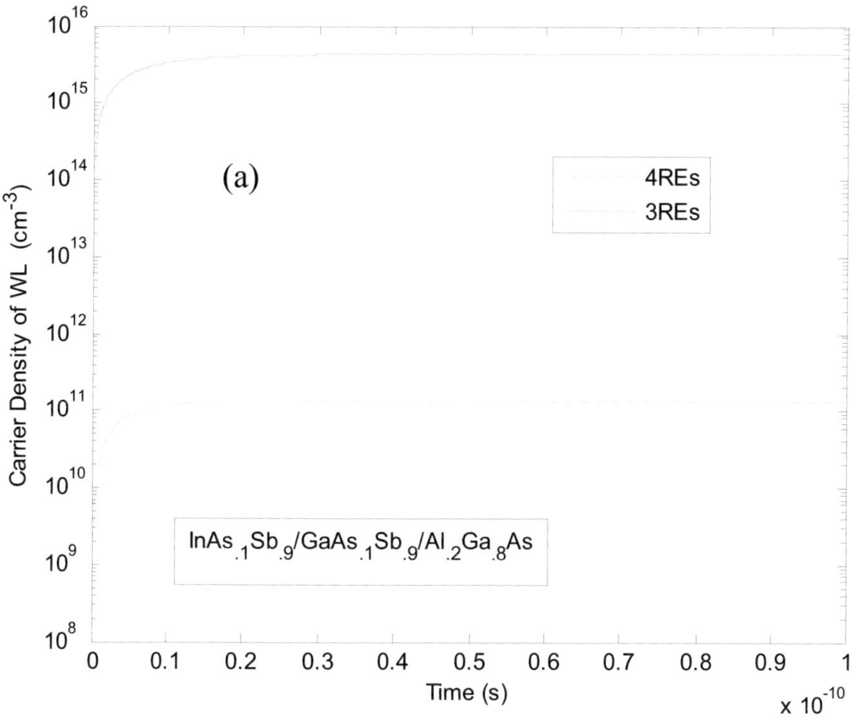

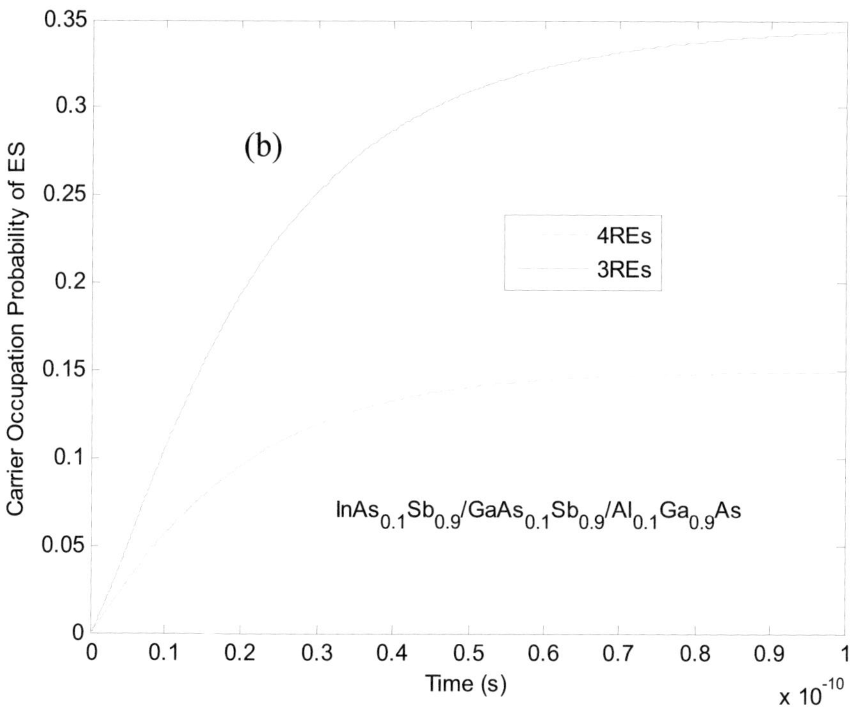

Fig. 8. (Continued)

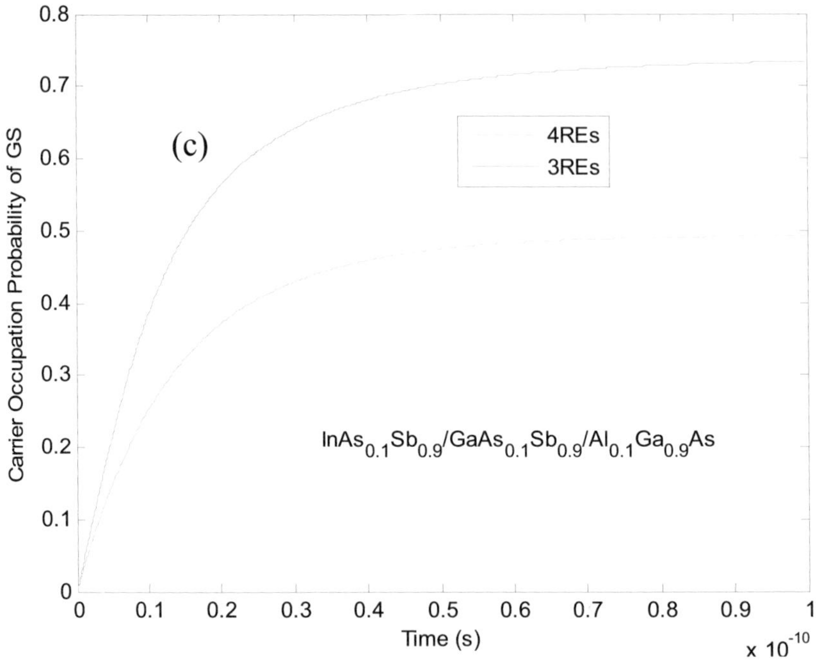

Fig. 8. A comparison between three- and four-REs system for (a) WL carrier density, (b) ES and (c) GS occupation probabilities for $InAs_{.1}Sb_{.9}/GaAs_{.1}Sb_{.9}/Al_{.1}Ga_{.9}As$ QD-SOA.

8. Conclusions

The progress towards SOAs, and then QD-SOAs, is reviewed. The importance of Sb-based structures is illuminated. To see the effect of QD-SOA layers, a REs for QD, WL and SCH layers are solved numerically. This make us specify gain-input power relation for Sb-based QD-SOAs and then examine layers dynamics. Both QD and SCH layers are shown to have an importance in the static characteristics due to their considerable effect in structure confinement. WL is shown to affect carrier dynamics especially in the ES and GS. Global quasi-Fermi energy (which included SCH, WL, QD ES and QD GS) is shown to be important in the explanation of results. At all cases, the inclusion of the SCH in the REs is shown to be essential.

9. Acknowledgment

We acknowledge our indebtedness to Prof. Mohammed Jasim Betti (Dept. of English, College of Education, Thi-Qar University) for proofreading the language of this work.

10. References

[1] Weng W. Chow and Stephen W. Koch, Semiconductor Laser: Fundamentals (Berlin, Springer, 1999), 1.

[2] Scott W. Corzine, Ran-Hong Yan and Larry A. Coldren, "Optical Gain in III-V Bulk and Quantum Well Semiconductors," in Quantum Well Lasers, ed. Peter S. Zory (New York, Academic Press, 1993), 18.

[3] Y. Ben Ezra, B. I. Lembrikov and M. Haridim, "Ultrafast all-optical processor based on quantum-dot semiconductor optical amplifiers," IEEE J Quantum Electronics 45, no. 1 (2009): 34.

[4] Michael J. Connelly, Semiconductor optical amplifier, (New York, Kluwer academic publishers, 2002), 3.

[5] Baqer Al-Nashy, Static and dynamic characteristics of Sb-based quantum-dot semiconductor optical amplifiers, a thesis, (Baghdad-Iraq, Baghdad University, 2010), 6.

[6] H. Shao and others, "Room Temperature InAsSb Photovoltaic Detectors for Mid-Infrared Applications," IEEE Phot. Tech. Lett., 18, no. 16 (2006): 1756.

[7] R. Sidhu, N. Duan, J. C. Campbell, and A. Holmes, "A Long-Wavelength Photodiode on InP Using Lattice-Matched GaInAs–GaAsSb Type-II Quantum Wells," IEEE Phot. Tech. Lett. 17, no. 2 (2005): 2715.

[8] G. Liu and Shun-Lien Chuang, "Modeling of Sb-based type-II quantum cascade lasers," Phys. Rev. B 65 (2002): 165220.

[9] Jungho Kim and others, "Static gain saturation model of quantum-dot semiconductor optical amplifiers," IEEE J. Quantum Electron. 44, no. 7 (2008): 658.

[10] Baqer Al-Nashy and Amin H. Al-Khursan, "Linear and nonlinear gain of Sb-based quantum-dot semiconductor optical amplifiers," Recent Patents on Electrical Engineering, 3, no. 3 (2010): 232.

[11] Y. Ben Ezra, B. I. Lembrikov, and M. Haridim, "Specific Features of XGM in QD-SOA," IEEE J. Quantum Electron. 43, no. 8 (2007): 730.

[12] Baqer Al-Nashy and Amin H. Al-Khursan, "Completely inhomogeneous density-matrix theory for quantum-dots", Optical and Quantum Electronics, 41, 989-995, 2010.

[13] Dieter Bimberg and others, "InGaAs-GaAs quantum dot lasers," IEEE J. Selected Top. Quantum Electronics 3, no. 2 (1997): 196.

[14] M. Vasileiadis and others, "Potential of InGaAs/GaAs quantum dots for applications in vertical cavity semiconductor optical amplifiers," IEEE J. Quantum Electronics 14, no. 4 (2008): 1180.

[15] D. W. Reschner, E. Gehrig and O. Hess, "Pulse amplification and spatio-spectral hole-burning in inhomogeneously broadened quantum-dot semiconductor optical amplifier," IEEE J. Quantum Electronics 45, no. 1 (2009): 21.

[16] Y. Ben-Ezra, B. I. Lembrikov, and M. Haridim, "Acceleration of gain recovery and dynamics of electrons in QD-SOA," IEEE J. Quantum Electronics 41, no. 10 (2005) 1268.

[17] Jin-Long Xiao and Yong-Zhen Huang, "Numerical Analysis of Gain Saturation, Noise Figure, and Carrier Distribution for Quantum-Dot Semiconductor-Optical Amplifiers," IEEE J. Quantum Electronics 44, no. 5 (2008): 448.

[18] P. Mock and others, "MOVPE grown self-assembled Sb-based quantum dots assessed by means of AFM and TEM," IEE Proc.-Optoelectronics 147, no. 3 (2000): 209.

[19] M. Mehta, A. Jallipalli, J. Tatebayashi, M. N. Kutty, A. Albrecht, G. Balakrishnan, L. R. Dawson, and D. L. Huffaker, "Room-Temperature Operation of Buffer-Free GaSb–AlGaSb Quantum-Well Diode Lasers Grown on a GaAs Platform Emitting at 1.65 µm," IEEE Phot. Tech. Lett. 19, no. 13 (2007) 1628.

[20] Amin H. Al-Khursan, M. Al-Khakan, K. Al-Mossawi, "Third-order non-linear susceptibility in a three-level QD system," Photonics and Nanostructures – Fundamentals and Applications 7, no. 3 (2009): 153.

[21] H. Al-Husseini, Amin H. Al-Khursan, S. Al-Dabagh, "III-nitride QD lasers", Open Nanosci. J. 3, (2009): 1.

[22] Wei Zhang and M. Razeghi, "Antimony-Based Materials forElectro-Optics," in Semiconductor nanostructures for optoelectronic applications, ed. Todd Steiner, (Boston, Artech House, 2004) 229.

[23] Shun-Lien Chuang, Physics of optoelectronic devices, (New York, J. Willy & sons, 1995), 708.